Sheep

ROAD

MAILBOX

COLD
ANTLER
FARM

gibson →

WATERING
HOLE

CHICKENS

Stream

S
W

DEC - - 2013

also by jenna woginrich

Barnheart

Chick Days

Made from Scratch

ONE
WOMAN FARM

3th   4th

21 1st   2 5 —   1st

22 1st   25 —   2nd

30 1st   26 —   1. verse

33 1st   28   1st

      27   1

      27   2

      26   2

      29   1st

ONE WOMAN FARM

MY LIFE SHARED
—WITH—
SHEEP, PIGS, CHICKENS, GOATS
·AND·A·
Fine Fiddle

JENNA WOGINRICH

illustrations
by Emma Dibben

Storey Publishing

The mission of Storey Publishing is to serve our customers by
publishing practical information that encourages
personal independence in harmony with the environment.

Edited by Deborah Burns and
Dale Evva Gelfland
Art direction and book design by
Carolyn Eckert

Illustrations by © Emma Dibben
Scanned objects by Carolyn Eckert
Author's photograph by
Tim Bronson
Lettering of title on cover by
© Mary Kate McDevitt

"Fall Is Here," quoted on
page 23. Words and music by
Charlie Maguire. Copyright
Mello-Jamin Music. All Rights
Reserved. Used by permission.

"More Wood," quoted on
page 47. Words and music
by Dillon Bustin. All Rights
Reserved. Used by permission.

"The Good in Living," quoted
on page 29. Words and music
by Steven Sellors. All Rights
Reserved. Used by permission.

The information in this
book is true and complete to
the best of our knowledge. All
recommendations are made
without guarantee on the
part of the author or Storey
Publishing. The author and
publisher disclaim any liability
in connection with the use of
this information.

Storey books are available
for special premium and
promotional uses and for
customized editions. For
further information, please call
1-800-793-9396.

STOREY PUBLISHING
210 MASS MoCA Way
North Adams, MA 01247
www.storey.com

Printed in the United States by
R.R. Donnelley
10 9 8 7 6 5 4 3 2 1

Library of Congress
Cataloging-in-Publication Data

Woginrich, Jenna.
   One-woman farm / by Jenna
Woginrich.
      p. cm.
   One woman farm
   ISBN 978-1-60342-718-0
(paper with flaps : alk. paper)
   ISBN 978-1-60342-867-5
(ebook)
   1. Farm life--New York
(State)—Washington County.
   2. Women farmers—New York
(State)—Washington County.
   I. Title. II. Title: One woman
farm.
S521.5N7W64 2013
630.9747'49—dc23

2013012525

TABLE OF CONTENTS

thank you

I'd like to thank all the family, farmers, readers, friends, and teachers I have had along the way. Especially my parents, Pat and Jack Woginrich, who are tolerating a farmer as a daughter — however exasperating and confusing it may be. Thank you, Brett McLeod, the best friend a farm gal could ever have. I must also express my intense gratitude to my Washington County neighbors Patty and Mark Wesner, the Daughton family, Tom Brazie, Jon Katz and Maria Wulf, and all the tribespeople down at Common Sense Farm. Thank you to Paul Fersen, Cider Dave, Ajay Rubin, Riding Right Farm, and all the others who have taught me on my journey. I really want to thank everyone at Storey Publishing, especially my editor Deb Burns, who helped form this book in its infancy, and designer Carolyn Eckert, who expressed it so well. Thank you to all my old friends at Orvis who watched me evolve from new office employee to full-time farmer. Thank you to Connie Brooks and Battenkill Books, as well as New York state's own Greenhorns, for cultivating a community where lingering woodsmoke can be the only truth that you have neighbors outside your own farm's boundaries.

And thank you to my pack: Jazz, Annie, and Gibson, who have never left my side.

AUTUMN   autumn   autumn   AUTUMN   AUTUMN
WINTER   winter   Winter   WINTER   WINTER
SUMMER   Summer   Summer   SUMMER   SUMMER
SPRING   Spring   Spring   SPRING   SPRING
HOLY OCTOBER   Holy October   HOLY OCTOBER   HOLY OCTOBER
JANUARY   FEBRUARY MARCH APRIL MAY JUNE
JULY AUGUST SEPTEMBER OCTOBER NOVEMBER
DECEMBER

DAYS OF GRACE

AUTUMN   Autumn
WINTER
SUMMER
SPRING
HOLY OCTOBER

Here's to the ones
who came before us.

The thing was to get now and then **elated**.

— FROM "IN A GLASS OF CIDER" BY ROBERT FROST

My friend Paul once ran a dairy of considerable size here in the upper Hudson Valley, but I met him when we worked for the same company, plugging away in a sanitized office. Paul always stood out from the rest of our coworkers, mostly because he looked exactly like I felt: so very out of place. When I think of him with his farmer's stride walking across those buzz-cut lawns of corporate carpeting, I smile.

On a wet, dreary, late-November day, Paul looked out at the bleak scene on the other side of our sealed office windows and sighed. Leaning back in his chair, he crossed his arms, smiling, then said softly that these were the Days of Grace. I asked him what he meant. He explained that the *Days* are what farmers in this area call the time of year between fall's fireworks and the first snowfall — when everything in nature is in a state of transition and naked waiting. This fragile period is a window of reverent preparation, a gift of last chances to farmers in our four-season climate to get everything done before winter nails us. The Days of Grace are filled with tasks like stacking wood and repairing tractors, loading the last hay and feed grain in barns, and oiling the snow blower so that it's ready to growl. Anything and everything necessary to prepare for the harsh season ahead gets completed, and everyone is grateful for the stolen time.

This becomes a chance for quiet reflection in a life that forces constant vigilance and resourcefulness. Weeds are long dead. Cash crops have been sold. Wallets are fatter. And mornings start a little later. The entire world takes on the calmer veil of the shoulder season.

Hearing Paul talk about his feral holiday, I felt like a child on Christmas Eve. Paul gave me the ability to see the holidays that were hiding out past the fence posts and swirling around milk cans in the corner of the barn. Something was waking up inside of me and bringing back a sense of lost tradition that I was missing dearly. It was a relief to have it fill up the dusty place between my lungs and heart again.

Days of Grace struck me, and struck hard. I wanted to know it; I wanted to be in that club. Farmers became a sacred tribe to me, and their November mornings were so much holier than mine. If not for the weather and pages on a calendar, it could have been any morning of the year for me — every Monday-through-Friday routine was precisely the same. People who farmed had a different way of understanding time, one based on sunlight and seasons, ebbing and flowing in activity like river water. Their year was alive, growing and dying. I felt cheated working indoors. I was so jealous it stung.

As those cold November days in the office plugged on, I imagined how it could feel to light the woodstove, pour a cup of coffee, and sit down with the newspaper, my dogs curled up at my feet. I desperately wanted this. I was grateful for my full-time job, but it was still a place I could not leave until darkness fell — a whole day lost to a computer and walls. I obsessed about the ranchers, corn growers, and dairy owners sitting around the table at the Wayside Country Store just a few miles away from my office. With their morning work behind them and the season ended, those men and women gathered over breakfast, talking about grain prices and silage and gossiping about their neighbors. They were free — and making life matter in a visceral way.

I ached to join their ranks.

COLD ANTLER FARM

When I first saw the place, I knew I had found my farm. The 900-square-foot house, built in 1850, was white and stout, sitting on the property's only level piece of land. Its center of gravity was a red door and three windows, no more complex than if you asked your 10-year-old nephew to draw a house for you with a ruler. It had a steep slate roof that was more than a hundred years my senior, and the kind of angle that made you think of ski lodges. I smiled up at it. A dramatic roof = a dry house. This place had good bones. And I liked the look of it.

All around it was six and a half acres of forest and rising land, so even though this outcrop was free of trees, it still felt protected and surrounded by the maples and oaks. A stream gurgled behind us, cutting a route through the snow and ice. A small artesian well bubbled to my left, under the grandest maple tree I had ever seen. Picturing her in autumn's glory, I got goose bumps. If hobbits had to live above ground, they would like it here.

To the house's right was a steep hill free of trees, save for a few old apples that twisted out of the snow as if Tim Burton had drawn them. Beyond were open fields, presumably an old pasture that the current residents mowed as an extended lawn. To the left was a barn, just as simple and quiet as the house, with red paint fading in the way people who reproduce "primitive" barn-board furniture try to replicate. Beside it was the old chicken coop, same red as the barn but with a fake chimney stove pipe jutting out the top. With the snow on it, it looked cozy. The place was free of livestock, but I could see its potential. I didn't even care what the inside of the house looked like. This was exactly what I needed: a small farm I could afford that made me feel safe. It's a good thing to feel before you step into a place you already love.

I acquired the place through a combination of luck and pure, unadulterated stubbornness. It took a few months of paying off credit cards, pinching pennies, and eating pasta every night, plus a kind and savvy real estate broker, a special USDA program, and a Hail Mary book deal. But I made an offer, and it was accepted.

I moved out of my old Vermont cabin that spring and became a resident of the Empire State. With me went two dogs, chickens, a pair of geese, three rabbits, and three sheep. I had big plans for a sheep-herding dog, more sheep, pigs, honey bees, and maybe even a horse I could ride along the winding road.

My part of Washington County is a mostly agricultural landscape based on dairy with occasional forests and sloping hills. There are a lot more cows than people, a lot more blue-collar jobs, and a lot more folks who spend as much time outside as those in Vermont — but they do because it's their occupation, not their recreation.

Jackson is a town of fewer than 700 households, sheltering some 1,700 people, positioned between the larger villages of Cambridge, Salem, and Greenwich (pronounced "*Green*-Witch"). It is part of the Upper Hudson Valley and home to fields, mountains, streams, rivers, and every sort of farm under the sun. Around my little freehold are cattle farms, alpaca farms, fish hatcheries, miniature horse trainers, artisanal cheese makers, and wood-fired bread businesses right alongside conventional corn, dairy, and soybean operations. It has historical roots in the American idealism of country living. Grandma Moses's farm is literally just down the road. I wasn't sure if I could live up to one of her paintings, but I was going to pull a hamstring trying.

And so my farm is where I'll follow this story of a year. A year of timeless work in honor of fellow agrarians, past and present, who have shared my climate and diet. This story will start and end in the same place — October — and weave through the work of woodstoves, cart horses, sheepdogs, gardens, orchards, livestock, and music. It's the story of these things and my learning about them here in Washington County, of days still kicking uphill here where oaks and maples sway brilliant red and gold under the cold October sun.

Every month has its own story, temperament, and celebration, and between the timelessness of the farm and that heady dance we call tradition, you will — as I do — see these months as living things you smell and touch. October is visited twice because that holy month is my anchor point — the four weeks of the year when my endorphins speed up, and every day I am six years old again. It's also the month I have the most to say about since so much living is packed inside it.

Becoming a farmer has turned calendar pages
into irrelevant symbols. Days of the week do not
matter like they used to. I'm damn sure a ewe trying
to deliver a lamb doesn't care about meeting a
spreadsheet deadline. The previously understood
calendar year has chosen to stop and turn around
on me like an obstinate workhorse. My year is now
measured not by days but by life cycles. My holy
days are based on the work and events of the farm
and of the seasons. I need no liturgical trappings or
pews, just a used Dodge pickup and some livestock.

I have discovered a wealth of ritual in this farm,
and it suits me. It is everywhere, in every part of
my day. My calendar is blessed with these annual
holidays — Apple Gathering and Lambing among
them — events I look forward to with the same
fervor I felt in my footie pajamas waiting for Santa.
The night before my first Shearing Day was wrought
with excitement and plans. I lay in bed feeling
(for the first time in nearly a decade) the innocent
excitement of a kid. I now was a part of the timeline
of human civilization, a part of an infallible
religion: Traditional Agriculture.

Becoming a farmer ushered me into the gospel
of dirt, life, sex, and death. My pillars of faith are
simple and yet timeless to the human animal.
I revel in them.

AUTUMN

I do not own a gym pass or see a therapist.

I'm not against these things;

I just have a farm instead.

It works for me.

NEED FIRES

There's an ancient tradition in the Scottish Highlands called *Tein'-éigin* (in English, the Need Fire). Whenever a group of farmers or clansmen fell into a particularly bad patch of luck — diseased cattle, looming war — the Need Fire was the remedy. The clan would extinguish individual hearth fires and start a new fire in an open space where the entire village could gather. This special fire was started not with a match or fuel but by friction. You needed to light embers with the traditional methods of rope against wood, because this was a blaze to be earned. This was the fire that invoked change and blessing, healing and calm, and kept the people strong.

Once it flamed high, they added wet wood to create smoke — lots and lots of smoke that filled the village and blew through people's hair. Farmers would run their cattle or horses through it as baptism and cleansing. The smoke was supposed to heal and bless all it touched. Then everyone grabbed coals from the common fire and took them home in small iron cauldrons to start their kitchen hearths anew. They lit those home fires from the mother coals of that ritual, knowing that the whole clan was there together in whatever happened. They had the proof in their hands.

This Need Fire is more than superstition: it's faith that through our directed efforts we can change our luck. The Need Fire is a connected people at collective prayer, an action on the soil itself, involving elemental basics of human survival tied with force and hope.

Perhaps it's the farmer in me, or the romantic, but I see the same hope in the steam curling off a coffee cup in a church basement AA meeting as in the smoke swirling from a 1356 Highlands bonfire. I know if my own clan up here in Washington County ever fell on hard times I would feel a lot more confident we'd get through it all together than if I had to deal with it alone. Strength comes from community and that hasn't changed, nor will it.

october

October himself is my *Tein'-éigin*. The entire month is ablaze with hard work, prayers, memories, and community. Here in the Upper Hudson Valley the leafy maples, ashes, and oaks explode into the colors of the fire. Reds, yellows, oranges, and all shades in-between light up the world for one last bonfire before those sacred Days of Grace begin.

Some people associate autumn with death, and in a way that is accurate. Winter is coming, and plants are fading into senescence. Leaves are falling off trees to feed the soil, leaving skeletal branches, naked and cold. But we all know those trees aren't dying, just hibernating. They spent all summer collecting sunlight, harvesting and storing their own food deep inside, and their leaves will return in the spring with a green so bright, we'll have to shield our eyes or go happily blind.

1 October. Anchor

At Cold Antler, October is my anchor. It's where
my year begins and ends because it's when the farm
work that governs this life begins and ends. The
garden has been bedded down and tomatoes line
the larder shelves inside Mason jar reliquaries.
Beans, pesto, and other greens sleep in the freezer
until they are brought back to life one snowy day.
The lambs have been weaned, sold, or are now old
enough to breed as Ram Time approaches.

When the hay and
straw are in the
barn

Stacked to the roof,
tier on tier,

Smelling like a
summer come and
gone

Around the place
I know that
Fall is here.

Hay is stocked in the old barn behind the
house. Although it's never as much as the
farm will need, it's as much as I can afford —
and at least enough to let me rest easy and
enjoy the month-long celebration that is
October.

That relief is the first gift of the year.
It makes me feel like a sleeping cat by a
woodstove: for the moment completely
content.

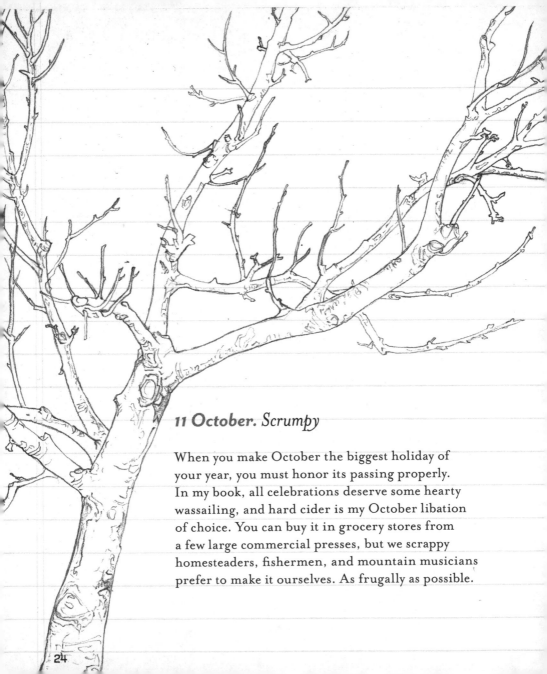

11 October. *Scrumpy*

When you make October the biggest holiday of
your year, you must honor its passing properly.
In my book, all celebrations deserve some hearty
wassailing, and hard cider is my October libation
of choice. You can buy it in grocery stores from
a few large commercial presses, but we scrappy
homesteaders, fishermen, and mountain musicians
prefer to make it ourselves. As frugally as possible.

A large group of us meet at an orchard where a "drop deal" has been struck. This means the Orchardist in Chief will let us fill up a pickup truck or two with "drops" (apples that have fallen off the trees and are usually enjoyed only by deer and other four-legged critters) for a ridiculously low fee — around forty dollars. Commercial orchards can't (or just don't) sell the apples that hit the ground, so we show up to clean house. It takes a few hours, but bucket load by bucket load we leave with a few hundred apples. They may be slightly bruised or gone off, but as long as they aren't black and soft, we consider them fine for our hooch.

Once we've loaded our orchard drops, we hit the back roads of Washington County and southwestern Vermont where we knock on doors, check with local officials, and plain old steal the wild apples we find on the roadsides or in the woods near our homes. These wild apples aren't as large as the orchard hybrids, but they are tart as can be. That combination of domesticated sweetness and

Scrump = to steal fruit

Scrimp = small apple

wild bite is what makes our signature blend of hard cider.

Hauling our scavenged goods, we make a caravan to Dave and Sue's house on the Vermont side of the border. Dave's modified antique press will reincarnate our apples as scrumpy. This form of hard cider is so dang strong, it has surpassed the name of cider, and ends up at around 12 to 14 percent alcohol by the time we get our lips around the bottles in early January. It has quite the kick to it.

Think of a flat apple champagne with a few shots of whiskey added to make it more interesting. Our scrumpy ages like wine, and as it ages, it gets stronger (if not in alcohol, then in legend).

Every **month** has its own story, temperament, and celebration.

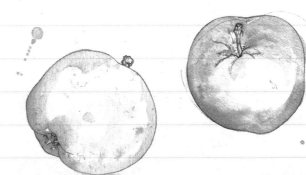

12 October.
Clan Isn't Always What You're Born Into

Your clan isn't necessarily your current friends, family, and acquaintances. You may have to seek outside your comfort zone for people who want to share the firelight with you. A tribe that supports and respects your dreams is a rarity — but if you can kindle it, you are rich.

Sometimes we get lucky and it is our own siblings, parents, childhood best friends, and cousins around those bonfires. Sometimes it isn't. A clan isn't necessarily what you're born into or marry. A clan is people who wrap you in their support and concern — people you can call at 1AM and they will come running with shovels. When you find them, you find home. They are out there, waiting.

Trick is you must look. You don't have to be religious to let the *Tein'-éigin* burn in your heart. You do need to accept that a better life is something worth believing in.

May your clan light the way.

14 October. *Pluck*

Fiddling is more about the season than the activity —
and nothing makes me reach for my fiddle like
October's crisp exhalations. I could be sawing out
tunes at an apple pressing or in the Agway parking lot.
Something about the color and the smells seems to
create the physical space and climate for Dorian and
D-major tunes, and I can't stop myself from playing.

I'm a self-taught musician who values stubbornness
over discipline — a card-carrying member of the
Determination Party. I didn't even know how to
string a violin, but by god, I wanted to be a fiddler,
so I was going to figure out how. I bought the
cheapest fiddle on eBay and a book called *Old Time
Fiddle for the Complete Ignoramus* (fitting) from some
instructor in North Carolina and taught my hermit
self to play.

I quickly learned that while a violin is a delicate
Arabian show horse, a fiddle is more of a sturdy
Haflinger. And this particular fiddle was a cart
horse! The book and accompanying CD were my
only teachers, and they let me be a little pluckier
than conventionally taught violinists could ever
be. I practiced with the dedication of the Recently
Inspired until I could hear fiddle tunes in movies
and television shows and readily copy them.

If the fiddle
strings felt no
bow-stroke,

If the concertina
bellows broke,

If no one sang or
cracked a joke

Then where's the
good in living?

— FROM "THE GOOD
IN LIVING"
BY STEVEN SELLORS

No. 149.

Autumn.

CLYDE WILLARD.

18 October. Cidering

The true work of making cider — or scrumpy — is in
the pressing. You need a machine that will take
whole, washed apples and crush them (using enough
force to collapse the hood of a car) into pulpy
chunks. Dave's modified antique press is a Civil War
cast-iron monster, a piece of junk he salvaged
from a New York orchard's estate sale and restored
into working order.

We split the work into several jobs, with constantly
rotating shifts and people. One person gets a large
bucket of apples out of the truck and dumps it into
the metal-grated bins for power washing. As the
power washers spray each new bucket of drops,
someone else is already hefting a bucket of just-
washed apples over to the press. There, one person
feeds the giant-toothed hand-cranked crusher
one apple at a time while a hard-working volunteer
turns the big iron wheel that grinds them up.
The work is hard to start, but once you hit your
stride you crank that puppy like you were born to it.

The last job in our assembly line is the signature
"presser." He turns a heavy wheel as well, but
this one lives on top of the press, separate from
the crushing wheel, where it twists a huge screw

with a wide, flat base down onto a pile of apple crushlings in burlap. The huge iron screw doesn't actually touch any apples; it presses all its force onto a giant round, wooden slab, like a barrel lid. As the wheel presses down, juice pours out of the apples inside the burlap and runs down a plastic-lined runway into a large bucket or carboy. The old timers call these crushed apples "the cheese" since they are squeezed out like a round of cheddar in a cheese press.

We work all day, without any real plan save for the unspoken rule that if you aren't doing something, you should be.

People slide in and out of roles, taking a break from pressing to power wash or to enjoy a plate of potluck food. Pulled pork, mac-and-cheese, and desserts line Dave's workshop as well as cups to enjoy a glass of fresh press with the food. Someone always brings dark beer and most of us are buzzed by early afternoon, laughing to the sounds of the turning wheels.

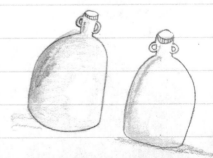

CIDER PRESS

O apple tree we'll wassail thee

And hoping thou will bear.

The Lord does know where we shall be

To be merry another year.

To blow well and to bear well and so merry let us be

Let everyone drink up your cup

Here's a health to the old apple tree.

— TRADITIONAL BRITISH
FOLK SONG

20 October. *Have You Herd?*

My Border Collie, Gibson, and I spend most of
the day outside with the flock, taking care of the
seasonal hoof trimming and body inspections.
I am with my flock every day but that doesn't mean
I flip them over and smell their hooves for rot.
Today I do.

The sheepdog is outside the woven wire fence,
having done the work of bringing all the wethers
and ewes to this pen near the farmhouse's front
door. His tongue is long and his mouth is wide
with a happy pant. The sheep are also a bit wound
up, but more relaxed with a good fence between
them and the wolf.

I grab a small Cotswold and turn her on her rump.
She's a little thinner than I would like and I decide
to drench her for worms as well as trim her feet. She
is passive against my bent legs, her head cradled near
my stomach as I do the work of an ovine pedicurist.
Gibson watches with pure envy.

26 October. *Fiddling*

On this Cidering Day I bring along Gibson and my
fiddle. Gibson spends most of the day exploring
the property or sniffing up friends' dogs also along
for the adventure. At one point after a particularly
rough go at the crushing wheel, I stop to take a swig
of Guinness and eat some slow-cooked pork while
I take in the scene: everyone in dusty flannels and
ripped jeans, everyone working hard, and everyone
smiling.

We're all in it for the scrumpy, sure, but we're also
in it for the tradition, which in this corner of the
Northeast goes back to before the Revolutionary

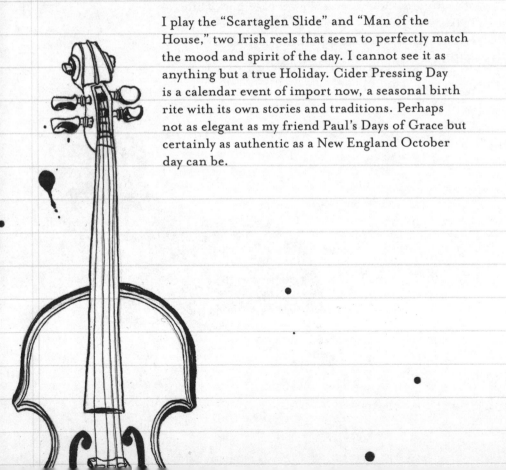

War. It feels good to be a part of it. And with my stomach full and the start of a buzz filling my head, I can't help but pull my fiddle out of my case and play a few rowdy tunes. Gibson, used to this clatter, curls up at my feet. A farm dog knows when to run and when to lie down.

I play the "Scartaglen Slide" and "Man of the House," two Irish reels that seem to perfectly match the mood and spirit of the day. I cannot see it as anything but a true Holiday. Cider Pressing Day is a calendar event of import now, a seasonal birth rite with its own stories and traditions. Perhaps not as elegant as my friend Paul's Days of Grace but certainly as authentic as a New England October day can be.

november

November mornings are a frantic place to spend
your time, at least in Washington County. The
parade is over. The skeletal trees hang around
like horror-film extras, and every day we expect
flurries and frost. And even if we're not made jumpy
by meteorologists, we don't let ourselves get too
comfortable either.

We use these Days of Grace to prepare for the
months ahead. The certainty of winter falls heavy,
and you can't drive three miles into town without
seeing dump trucks full of firewood, sliding past
the fuel-oil truck chuggin' toward you from the
opposite direction. No one is coasting. Farmers
either use every moment to finish those sacred
last-minute tasks before snowfly or they've already
sealed the envelope on the season, far more
concerned now with deer camp arrangements.

1 *November. Samhain*

Back in the eighth century, Pope Gregory chose
November 1 for All Saints Day's because the
Celts were then already observing a fire festival
known as Samhain (pronounced *sow-en*, Gaelic
for summer's end).

Lots of rumors and mythology swirl around the
old holiday, but Samhain was not about sacrificial
goats or satanic rituals. It honored family members
who had died during the harvest year. It was a time
to take inventory of seeds, herbs, and grain stores
for the long winter ahead. Perhaps it developed its
current dark mystique because it said goodbye to the
daylight half of the year and marked the beginning
of the darker half. And with the work of growing the
winter's food done, there was finally time for both
reflection and sorrow.

Samhain is a quiet day here. I think of the people I
lost, through death or the entropy of lives moving in
different directions. At sunset I light a bright white
candle and spend a little time remembering them.
It's a humble ritual, but one that keeps me grounded
around all the harvest parties and celebrations. It
carries me through.

3 November. *Winter Pig*

Back in Vermont I started the tradition of the Winter Pig. In late fall, after Hallows, I buy a feeder pig or two. This year, ten-week-old Pig will live in the barn until she is ready to be harvested in February.

That may seem like an odd time of year to be putting pork in the freezer, but for this small farm it makes perfect sense. Winter means that the pen in the corner of my barn is safe and comfortable for Pig. In summer it would be a hot, smelly mess. Most people raise pork on pasture, but the precious little pasture and wood space I have goes to animals not raised for slaughter, like horses and wool sheep. Also, around here pigs get loose all the time and wander into the woods and neighboring farms, causing damage and accidents a-plenty.

Raising Pig in winter keeps down the hassle and smell, and since a comfort-loving pig is more likely to eat and nap in the barn than escape, she grows fast and true. And when Valentine's Day hits around this farm, my thoughts are red — but not about paper hearts. I am looking up recipes for how to cook real ones. Take that, Martha Stewart.

8 November. *Bun Baker*

My first winter as a smallholder in Vermont,
it dropped below zero for weeks. I kept having
to call the nice people at the fuel oil company
to deliver more of their sludge into my house.
I was spending more than $400 a month just
to keep the thermostat at 50 degrees. Jazz and
Annie, my two aging huskies, were welcomed
in the quilt-covered bed with my sheepdog.
Yes, that winter I learned what a "three-dog
night" really means.

Much as I loved my dog pile, I was ready
to add some alternative heating to my
farmhouse, and I opted for the traditional
firewood-burning woodstove. I want to be out
in the forest harvesting trees off my own land
and using my own horse to pull them to the
chopping blocks.

My Vermont Bun Baker is a winter farmer's
dream. I loaded it up with wood first
thing this miserable November morning
and set my big steel percolator with the
heavy bottom on top of it, rancher
style. There's room for a kettle of hot

Wood is held
sacred in many
cultures, but
few as much as
the Celts. Oak,
Ash, and Thorn
were the holy
trio of woods
to the Druids,
and the word
Druid actually
derives from
the Gaelic *Dru*,
meaning oak.

Locust makes the
finest fence
posts, and a
birch sapling
twig hung on
the front door
protects from
a storm. The
practical and
the traditional
thrive here.

water too. When I finish haying the sheep and horse and refilling chicken and rabbit feed and water containers, I come inside to a warm house, a hot cup of coffee, and a steaming teapot ready to pour over a simple breakfast of oatmeal.

The fire keeps the farmer going, and thus the farm. I can tend to them, all of them — from the sleeping sheep on the hill to the chickens on their windproof roosts. It is a comfort on cold nights to know it has been that way since the Druids prayed so long ago and far away.

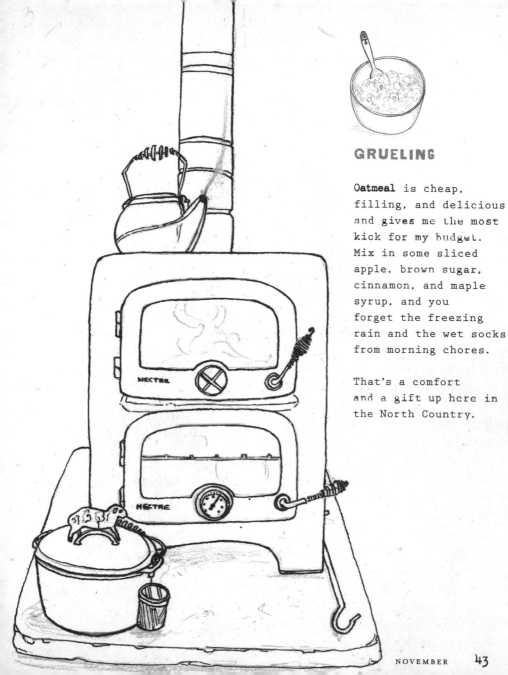

GRUELING

Oatmeal is cheap,
filling, and delicious
and gives me the most
kick for my budget.
Mix in some sliced
apple, brown sugar,
cinnamon, and maple
syrup, and you
forget the freezing
rain and the wet socks
from morning chores.

That's a comfort
and a gift up here in
the North Country.

12 November. Why Sheep?

The more involved with the animals I become,
back in Vermont and here, the more I realize that
sheep are the perfect small-farmer's investment.
They produce not only meat but milk, wool,
cheese, shearling, leather, lambs, and lanolin
as well. They are easy to manage, respectful of
fences, and small enough that if you get attacked
by one you probably won't die. And as a single
woman I can manage an entire flock with the very
green energy that is my sheepdog, which would
not be the case with pigs or cows.

The idea that I can get all this *and* work beside my
Border Collie is why I raise these animals. On a
bitterly cold winter night I can bolster myself with
a full stomach of lamb chops, cover up in a wool
hat, sweater, and sheepskin gloves, and feed my
sheep with just a few flakes of hay. I ask you, what
other animal gives us so much and asks so little
in return? And does it without jumping fences or
making a fuss?

18 November. Woodpile

November for me isn't about last-minute preparations, and that isn't through self-righteous Ant vs. Grasshopperism as much as from the nature of the farm. I'm a single woman managing land and livestock alone. I also depend on wood for 100 percent of my heat now, so unless I want to go into winter terrified, regretful, and cold, firewood better be in. Thus I have a mighty stack, at least four cords, arranged on the overhanging porch built into the side of the farmhouse.

My firewood pile is a mix of bought, bartered, and home-harvested. My friend Brett, a fellow homesteader and a lumberjack, took down some poor and dying locust trees out behind the barn and left them to harden and season, propped above the ground so they wouldn't decompose. When they were fuel-ready, they were pulled up to the woodpile by Jasper, my working pony, my homestead tractor. He's only 11 hands and 450 pounds, but in harness he is a force to be reckoned with. Together we pulled trees out of that woodlot at which even the most skeptical farmer would nod approvingly.

Well the sassafras it burns too fast,

It starts the fire but never lasts

And swamp oak likes to smoke you blow
it till you think you'll choke.

But hickory is just the tree to
remind you of the ecstasy

Of having a pile of good wood.

— FROM "MORE WOOD" BY DILLON BUSTIN

measures 4 feet
high by 8 feet long
and the depth
of the length of the
wood (usually 16 to
18 inches long, to
fit inside a wood-
stove or a fireplace).
A full cord, by
contrast, is 4 feet
deep.

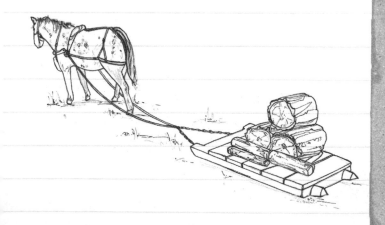

22 November. *Gratitude*

When I became a homesteader on a mountain in New York state, not only did I move hundreds of miles from my hometown in Pennsylvania, but I also moved a million miles from my middle-class-family upbringing — a choice that still confounds and disappoints my parents.

My new lifestyle with a farm full of animals and obligations means travel is impossible. So when first Thanksgiving and then Christmas roll around, choices like not going home for these traditional holidays have become wounds. My new life means a life without them. I explain that having a farmer as a daughter is like having a daughter in the army — one deployed indefinitely. People who choose to dedicate their lives to the land and animals, to grow food, to do *any level* of homesteading, learn quickly that company must come to them. In some families, this is a great divide. I invite mine to join me and I hope they will.

No matter how hard I look, my old path is gone. As a farmer in love with her farm, I only know that to pursue this love, this path, feels correct.

No. 11. The Dear Old Farm.

FRANK YELLAND.

JAMES L. ORR.

1. I love my home a-mong the hills, Where meads and brook-lets charm;
2. What sweet in-spir-ing joys a-bound, Free from all taint of harm;
3. How peace-ful-ly thy day-light's close When twi-light's cur-tains fall;

How rich and pure the bliss that gilds A life up-on the farm.
What hap-py mem'ries clus-ter 'round Thy hearth, thou dear old farm.
How calm-ly sweet is thy re-pose When dark-ness cov-ers all.

CHORUS.

I love the good old farm...... The dear, old, peace-ful farm;......
The good old farm, I love.................. the dear old farm;

Its fields are green, and its skies se-rene, I love the dear old farm.

No. 12. Work, for the Night is Coming.

1. Work, for the night is com-ing, Work thro' the morning hours; Work while the day is
2. Work, for the night is com-ing, Work thro' the sun-ny noon; Fill brightest hours with
3. Work, for the night is com-ing, Un-der the sun-set skies; While their bright tints are

30 November. *Let It Snow*

Late November pleads for change inside me. With
fall now a memory and winter one snowfall away,
I am addicted to the weather radio. I crave snow the
way a sinner craves a clean slate. It's not so much
that it creates beauty but that it validates my need
to be present here. True winter commands my
constant vigilance, and I can let go of other people's
guilt-inducing ideas about me being anywhere else
through the holidays. My farm needs me. In winter,
I am the farm's soldier. This brings me peace
of mind.

Something about that first snow lets me exhale,
lets me accept that I am living on an upstate New
York farm, where weather is more real than any
family drama. When the farm is covered in snow,
the real chores come — raking roofs, defrosting
livestock tanks, hauling hay bales on sleds for
feeding, bringing in wood to keep the home fires
burning. It is in this work that I truly understand
the necessity, duty, and responsibility of my choice
to live on this farm.

Southbound Canada geese pass overhead, calling
out, and my own geese answer. This is my life now.

WINTER

LIGHT

Though a calendar wouldn't swear to it until
December 21, winter is here. It has turned around
three times, made itself comfortable, and lain
down at my feet.

The wood, seasoned, split, and stacked all summer
and fall, awaits its final destiny. Knowing it is there
is comforting. When I wake up on any winter day,
the fuel is the first thought in my mind. Nothing
else happens on this farm until that first match
strikes in the chilly morning air and lights the fire
that welcomes the day.

I hunch down in front of the woodstove in the
corner of my living room, knees bent, my rump
almost touching the ground. I feel like some
benevolent house fairy, working the magic that
will soon fill the farmhouse with warmth. There
in my crouch, the lit match in my hand, I open
the heavy iron and glass doors of the stove and place
some new tinder on last night's still-warm coals.

It's still dark, both outside and in, but once I start
a fire, the farmhouse is lighter, both in mood
and visibility. Since I have no overhead lighting,
save in the kitchen, I like welcoming a winter's
day like this, knowing that the first light that enters
my morning I personally worked to achieve.

Gibson has been by my side this whole time, his
black fur shining in the firelight. He's so soft, softer
than a working dog should ever be. I run my hands
over his back and thump his ribs. His tail drums
against the old floor, and we both know it is time to
face the work waiting for us outdoors.

Before I get dressed, I fill the percolator from
the crock and set it on the now-churning stove.
Tackling the frozen water, feed bags, hay, and
wind chill is easier when you know you'll return
to a warm shelter and the promise of comfort after
deprivation. This is the oldest song we know.

coffee

Announced by all the trumpets of the sky,

Arrives the snow, and, driving o'er the fields,

Seems nowhere to alight: the whited air

Hides hills and woods, the river, and the heaven,

And veils the farm-house at the garden's end.

The sled and traveller stopped, the courier's feet

Delayed, all friends shut out, the housemates sit

Around the radiant fireplace, enclosed

In a tumultuous privacy of storm.

— FROM "THE SNOW-STORM"
BY RALPH WALDO EMERSON

december

The first true snowfall of the season seems to
hit in mid-December every year. We'll get the
occasional flirtation or two before, an inch here
or there, but the snow that comes and stays until
March seems to fall around now. When it does
I am ready, and I welcome it with open arms. I spent
the summer and fall preparing for it and by now
there is hay in the barn and wood stacked dry and
waiting under the house. The snow comes as a happy
guard over the farm, keeping watch over all who
wander outside while we inside the farmhouse read
by the fire.

2 December. *Layered*

Gibson can run out into the cold morning wearing nothing but his black-and-white fur coat. I'm dressed in layers — multitudes of them. I wear flannel-lined jeans with heavy, hand-knit wool socks under rubber boots. From the waist up, my first layer is a cotton Henley, then a lightweight wool sweater, and over that a thick canvas vest. My hat and hands are covered in wool as well, gloves and a hat I knit from my own sheep's wool.

This is my uniform for winter chores, warm and yet flexible enough that I can open up bags of corn or grain. Chores do not take long, but I can focus on the work and animals the entire time and not worry about discomfort.

At the other end of the day, by the time dinner is ready, my clothing is strewn around me on the kitchen floor, wool sweaters with arms akimbo, looking like casualties of an apparel war. When the house is warmed up, all I need is a long-sleeved base layer to feel at ease. Candles and oil lamps provide the majority of the light. Although I have electricity, I prefer the flames' whispers and crackles to the grid's shout.

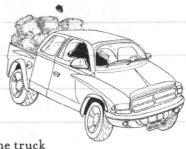

7 December. *Town Trip*

The farm always needs something, so the truck gets started up, rumblin' and ready to take on the mountain. Gibson jumps up to sit shotgun as we head into town. The local station out of Albany is calling for snow showers this afternoon, but I don't need the weatherman to tell me that. Both Gibson and I can feel it, the coming snow causing our skin to hum and the air to change and swirl.

Because the daylight is so dear and the energy is building toward Yuletide, the days seem short and powerful. When I stop into a store I'm met by celebration. Gibson accompanies me everywhere. The hardware store, feed stores, even my bank lets him walk in with me.

When we return from town, the fires are out and the house is cold. I carry my grocery bag in one arm and with the other I grab a couple of logs and bring them inside. I set a pot of water on the wood stove for hot chocolate, as the first snowflakes start to fall. Might as well do this right.

10 December. *Make a Wish*

For a year now I've lived with a small carting
pony named Jasper. He is a Pony of the Americas
or POA, a small Appaloosa-derived breed usually
used for children getting started in Western riding.
At only around 450 pounds and 11.2 hands, he's
too small for me to ride but perfect for harnessing
up and moving logs to split or a cart of manure
to spread. I love him, but I yearn for a riding horse
to be my partner in adventure.

My wish is for a Fell pony, a breed native to
northern England and Scotland.

I first came across this breed in an encyclopedia
of horses in a bookstore. Fells are built like
Clydesdales, all muscle, mane, and feathered feet,
but nowhere near as large. The one in the book was
as black as coal dust, his mane cresting along his
strong neck and sweeping down below his elbows.
He looked like something a hobbit would ride to
Mordor.

Since there are fewer than 200 Fell ponies in America, making them available only to wealthy hobbyists (and hobbits), the likelihood of getting my Christmas wish is on the order of running into a unicorn. But as long as I can picture his face, I won't give up my dream that my Fell is out there, somewhere.

So I make a wish.

14 December. *Presence*

December will always be a big party to me, but to the farm it is an altogether different story. Christmas can't hold back snowstorms, freezing rain, frozen gutters, high winds, and power outages. If I were to leave the farm and bad weather hit, an animal got sick or hurt, or someone froze to death or starved, I would never forgive myself.

I've chosen a life that requires presence in an almost monastic sense. Most mornings I get up at 4:45 AM and start the day feeding chickens and rabbits, then lugging bucket after bucket (some 300 pounds' worth) of water from the artesian well (there are no outdoor hoses), and hauling hay bales and 50-pound feed bags.

Any snow must be removed from the roofs of old barns and animal sheds to prevent collapse. This means walking around 6.5 acres with a roof rake and wiping down two sheep sheds, a horse pole barn, the old red barn, and the house's lower roof. When this is done, there is, of course, more wood to bring in, fires to tend, and the first meal of the day to see to so that I have the energy to do it all again in late afternoon. Even if I could afford

to hire people to watch over the farm with this many animals and chores, I simply wouldn't feel comfortable leaving my homestead in another person's hands.

And of course, I don't *want* to leave my farm. I wake up and the world shines! In a roaring thunderstorm or raging snow squall, I get out of bed and put on my rain or snow gear as if it were armor. I step out into the fray, crook in my wool-mittened hands, and I do the work I was meant to do.

15 December. Mail-Order Spring

When I come inside from morning chores, there is about a 30-minute warm-up in the living room. I pour coffee and get out my list for the day.

Today I do something extra. I pull out the seed and chicken catalogs and start planning my spring chick order, like a child browsing through the Sears–Roebuck catalog, page after page a wide-eyed wish list.

I have a few favorite breeds but I always keep an open mind to try something fancy. Will this be the year of Polish crested hens or long-tailed Japanese roosters on the fence posts?

I forget the cold outside. A farmer has to be a telescope, looking a season ahead while weathering the present. My fingers flip the pages and I smile, sipping the hot coffee. Already it feels like spring.

18 December. Yuletide

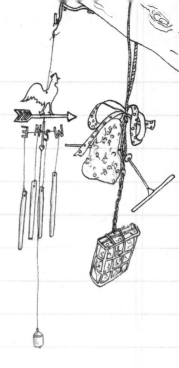

I cut a small tree and set it up in the farmhouse's front window, decorated with silver balls and black crows, tying an antler to the top. This is, after all, Cold Antler Farm, and I have my own symbols and ideas about what denotes love, luck, safety, and holiday spirit.

My December rituals now involve baking and planning the perfect holiday meal while listening to the Celtic Christmas channel on Internet radio. I knit gifts for friends and mail cards with images of snowy Border Collies and lambs. I bake pies in my living room woodstove for friends and deliver them in baskets.

One of my favorite post-chore rituals for Yuletide week is to read a chapter or two a night of Dickens' *The Christmas Carol* — a book I knew well from countless movies and cartoons but have never actually read till now. Christmas with Dick, I call it.

19 December. *Stoked*

In October, and even late November, you can start a fire in the evenings to fight off the chill, but the effort is basically ceremonial, not truly necessary. At that point, the temperatures will never drop below 15 degrees and scare the pipes into freezing. But December is a different ballgame, and if you want to wake up to running water and a toilet without a layer of ice in the bowl, you'd best stoke that fire late into the night and again early in the morning.

Come late afternoon, I need to relight the fires again — a repeat of the morning's ritual. Light fires, feed animals, feed self. I do these things in said order every winter day. It is as much a part of me as any religious ceremony is to a priest.

Ritual can be mundane. This does not hinder its import — at least not to this shepherd.

20 December. *Comfort Food*

At suppertime I usually heat up leftovers or cook
a simple meal like warm soup and crusty bread, the
kind of stuff I can dip into a mug and lazily enjoy
while I read or watch a favorite DVD. The dogs
and King George, the Maine coon cat, sprawl and
stretch, and I know as the bread hits my lips that
every animal in my keep is at least as comfortable
as I am.

The sheep and horse have bedding and hay and
shelter from the wind. The chickens are on the
roost in warm rows of feathers and measured
breaths, and the younger sheep watch them,
deerlike, legs tucked under and ears perked. Pig
has burrowed into her hay pile, curled up and
unseen. The snow starts to fall, coating the ragged
dead grass around the lamppost outside. I grab
a second helping of soup, tuck into my quilts, and
consider myself blessed.

21 December. *Longest Night*

The Winter Solstice is a night that commands
the attention of agrarians. Here in the North
Country, on the eastern side of a mountain,
it seems that every winter day starts to grow dark
by 3 PM, but come the solstice proper you feel
it in your bones.

It's not a time of despair, though, but of
celebration. The farmhouse is lit with lanterns
and candles and a crackling fire in the wood stove.
With fiddle music and a good home brew, it's hard
to focus on the dark. I just make my own light.

Most of us notice the change only when walking
to our cars after work: suddenly in January we can
see our keys in our hands again and are driving in
sunsets when it was dark as midnight a few weeks
earlier. But for those of us who live with livestock,
different clocks are ticking away the dark. As the
days grow longer, hens lay eggs again, sheep and
goats know their birthing time is closer, and owls
in the forest fly home late in the morning from
their mating rituals.

Come the Solstice I am both grateful and excited.
When I can gather friends, we light a bonfire on
that dark night and spend time around it, outside
together. We warm ourselves with scrumpy and
whiskey, fiddles and guitars, and we welcome the
return of the sun.

There's no family. No boyfriend.
Just a girl and her wolf pack and her
farm full of animals.
It's perfect and imperfect. Full and empty.
Validating and terrifying.
Above all, it is just as I want it to be.

24 December. *Farm Christmas*

Tonight, Christmas Eve, I sit in my home, talking
on the phone with friends in the glow of a twinkle-
light tree full of those black-winged angels I adore.
I pull a happy apple pie out of the stove and set it up
high on a counter to cool. It isn't snowing outside
but the sifting of powdered sugar covering up the
lattice crust will do for me.

I sing along with "Auld Lang Syne" and ignore the
drooling dogs at my feet. I tell them it is just celery
and to chill out. When it comes to Christmas Pies
I have no problem lying to my dogs.

So this is my Christmas. There's no family. No
boyfriend. Just a girl and her wolf pack and her farm
full of animals. It's perfect and imperfect. Full and
empty. Validating and terrifying. Above all, it is just
as I want it to be.

31 December. *My Life, My Love*

Far above money, location, or determination, I
have found, becoming a farmer requires one thing.
And that one thing will win you anything you want
in your future, but it will also rip you apart: Love.

I have been told I write far too romantically about
my life here at Cold Antler. That's because this
is a romance. I am head over heels for this place.
I love it and all its many imperfections, griefs,
and complexities. I love the beautiful things, like
a snow-covered farm and the newborn lambs in
spring, and I love the messy things, like strained
relationships. I even love the goose shit on my front
step because a life without goose shit would mean a
life without geese, which is beyond comprehension
at this point. I mean, how do people without geese
even know when their mail arrives?

And I strongly believe that loving this lifestyle —
whether it's a goat in your Seattle backyard or
40 acres in the Virginia backwoods — is the only way
to be successful. And by successful, I mean living a
life that makes you happy, surrounds you with good
meals, and builds community and a sense of place.
If you are truly in love with the idea of producing

your own food and caring for your own livestock, then it will happen because it simply must. You won't be content until then; you'll give up what you have to give up. You'll take the leaps and risks that you need to take, and you'll sweat and work until you can't see straight or feel your hands through the calluses.

You'll do it because it sustains you. You'll do it because the lack of it will eat you up.

january

January comes and so does a long-anticipated sigh. The holiday parties and festivities are behind me and I can finally stretch from head to toe and take on the wooly animal that is deep winter. I am going through firewood as if my mighty sticks of locust and maple are balsa planes and kite frames.

Jasper's once-sleek coat looks so scruffy and thick, I forget he is a pony. He looks like something out of a book of Ice Age equines.

The chickens are the true soldiers of the winter. They hunker down in their windproof bunker, sidling up to one another on the roost for warmth.

If a comb shows signs of frostbite, I rub on a little Vaseline to protect the exposed red crown from the worst of it. Despite this, every year some poor boy meets spring with a black-tipped comb. I always tell him there are worst fates for chickens. I'm not sure he believes me.

No matter how hard I look, my old path is gone.
As a farmer in love with her farm, I only know
that to pursue this love, this path, feels correct.

1 January. Plan A

People always ask me what my Plan B is if the farm
fails or people stop reading my books and blog. I
tell them what my good friend Ajay always told me.
Every time Ajay moved to a new state or town, he
cashed out of everything — all his possessions and
connections to old jobs — and focused entirely
on his new home city. He said that was what Cortez
had done: landing in the New World, he burned
his ships. His crew was highly motivated.

It's harder to fail when success is your only choice
for survival. At this point, my Old World is so far
away, so foreign to what my life has become, I do
not know if I could turn back.

I burned my ships, too. Here's to the New World.

5 January. *Books and Horses*

Some sixty people have come to Battenkill Books
this afternoon to hear me read from my book
Barnheart. Afterward, a fiftyish woman comes up
to shake hands, introducing herself as Patty. She
invites me to her farm sometime soon to go for
a cart ride with her road-safe horse, Steele, a
nine-year-old dappled Percheron gelding. I have
never agreed to anything with more certainty.

I have always planned to have horses be part of
my farming life, and I already own Jasper. I love
the little imp, but he is too small to ride for any
real length of time, so his job is sheep protection
and the occasional small carting jobs around the
farm. Jasper is my equine training wheels. We do
what we can, but without a mentor, and only books
and videos for study, we don't work as much as I
would like.

Now here is a woman who lives just a few miles
away, welcoming me over to learn under her wings.
The universe has presented me with a mentor,
and I have eagerly accepted.

7 January. *The Lamb Program*

The breeding ram, Atlas, is outside right now with
the flock. He'll stay here two months and in that
time he should have performed all the duties a ram
should. This means lambs on the ground in late
May or early June, while my other ewe, Martha, is
due in March.

TWO REASONS FOR LATER LAMBING

1. To make sure there is plenty of grass available for the flock.

2. To make it easier on me, the shepherd.

A later lambing date means the sheep I currently have can be rotationally grazed on the existing pasture as well as help clear land for new pasture in the woods. Since I do not have any specific market dates to meet, I can raise the sheep when I please. It'll be a lot easier on me and the mothers when the days are longer, the weather is comfortable, and the ewes will be shorn.

The downside is that by the time the lambs are eating a diet of mostly grass, we'll be well into fall. That means it will cost more to feed and fatten them through the winter on hay, minerals, and grain.

It's a trade-off. I do hope to sell or trade most of them right after weaning and by autumn keep just a handful of the sturdiest.

I find sheep to be worth many times their weight in firewood, lumber, farm services, chimney sweeping, and other things I can barter for. Lambs are a prime currency in these parts, so I aim to raise more than my seven ewes can offer.

11 January. Salad Days

Danvers 126 half-long carrots, dwarf Siberian
kale, Parisienne round carrots, deer tongue, and
speckled trout lettuce: those are the five crops I
have planted in my kitchen right now. My first little
yogurt container of kale is sprouting and looking
healthy. The rest are sitting under a little plastic
greenhouse on top of a heated mat. The seeds, the
greenhouse, and the heating mat all cost less than
fifty dollars, and I will use that greenhouse all
spring long to start seeds.

When I transplant this bunch of plants into the
garden beds, they will be protected under plastic

tunnels (tents, really) in the earliest outdoor mini greenhouses. Kale, lettuce, and carrots are hardy creatures and can handle an early season with a little babying. As soon as they are outside, another round of seeds will be started inside — broccoli, parsnips, and peas — and then they too will move, when ready, to a second series of covered houses.

By the time the real outdoor planting season starts, I will already have food waiting to be eaten and harvested. And instead of spending that time sowing peas and lettuce outdoors, I can use it to build new raised beds and put up critter fences.

I have big plans for the garden this year. This will be the Summer of the Salad.

All this time I thought
I was becoming a farmer, but instead,
I find the farm is becoming me.

13 January. *Ignition*

If I was wise last night, a pile of twigs, birch, locust bark, and small hatchet-sliced stove wood sits beside the stove in the morning, ready to be ignited. If I wasn't, with a heavy sigh I light a pile of wadded-up paper and coat it with splinters and bark shards nesting on the bottom of the metal wood caddy. Then I hurry to the cold mudroom behind the kitchen, to my indoor stash of dry wood and my little Fiskar's hatchet.

I chop into a piece of cordwood, grateful for the sharp blade. In no time I have a handful of slices of pine or birch ready to kindle into a proper blaze. It doesn't take long. When the sticks are burning well, I slowly add larger pieces, encouraging them with more paper or quick-burning bark.

I sit there and, like the opening sequence of a favorite television show, watch the fire roll through the credits of the endeavor. *Kindling: brought to you by foresight! Flames: starring birch bark and locust hulls! Also starring: pine shards and stove wood from special guest Finnish hatchet!*

14 January. *Herding Lesson*

Gibson and I are still beginners, still at the very
start of our training as shepherd and sheepdog.
It's not a race for us, and I am treating it as a
lifelong education of farmer and farm dog. When
the weather turns better, we'll get back to lessons
on a regular basis.

Today I want to give him a brush-up, so out we go
into the waning light and 18 degrees of a winter
afternoon. The work is a flurry of excitement since
Gibson has not been out with the sheep in weeks.
My farm is small, and the pasture only around
3 acres.

The sheep are out of sight, over a rise, and the
little 50-pound dog is shaking with anticipation.
A proper sheepdogger would have all sorts of
commands and ideas about the best way to gather
the wanderers but I just point toward the rise and
say, "Go get 'em." He is off like a buckshot.

The frozen ground cuts his paws, but he doesn't
notice. When we come inside I notice the red paw
prints on the farmhouse floor and call him to me.
I wash them carefully and give him some water.
He is soon asleep in his crate with bandaged paws.

19 January. *First Dance with Steele*

I'm standing in my friend Patty's driveway, under the
winter's iron sky, next to an 1,800-pound Percheron
that makes Jasper look the size of a white-tailed deer.
At 17 hands, his shoulders are above my head. A
flowing mane, gray and crinkled, hangs over his giant
head and quiet, watchful eyes. Despite his bulging
legs and chest and hooves the size of dinner plates, he
is as calm as a parked freight car.

Patty emerges from her farmhouse carrying a giant
leather collar over her shoulder and 16 feet of leather
lines. She explains what is what and what goes where.
I am learning the language the way a child learns it:
by holding up things and repeating what they are out
loud. *Carriage whip. Blinders. Singletree. Lines.*

We then pull up the large wooden cart behind Steele. A pair of long wooden shafts slide along Steele's sides, through loops in his harness, to direct the cart and keep it level. Patty climbs into the cart and pats the seat in invitation. I practically teleport beside her.

We amble down the dirt driveway, then start to trot as we hit the paved road, gliding across the landscape to the musical *clip-clop* of Steele's hooves. Grinning uncontrollably, I sit back in my seat and watch. There is no autopilot here. Patty constantly converses with Steele through the black leather lines, her voice, and her carriage whip. It is a dance, not a taxi ride.

Patty abruptly hands me the lines. It doesn't take long for my inexperience to have Steele smack in the middle of the road. Patty shows me how to use the lines like a steering wheel, giving and taking slack, watching Steele's feet and ears. In a few hundred yards I am better able to work with my dance partner, and my heart is beating its own tattoo along with his hooves.

This horse - THIS HORSE! - is a thing to behold. And in the next few moments we will get a 60-pound harness and collar on this massive beast, then hook him up to a big wooden cart. I am chomping at the bit before Steele even has one in his mouth.

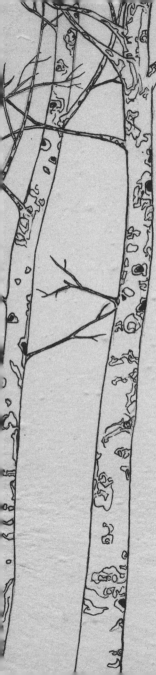

february

February is when the big storms hit and the farm
takes the worst beating of the year. Ice clings
to power lines and usually tears them down.
Blackouts are common enough, and every time
my world goes dark I am grateful for my lanterns,
candles, and those amazing woodstoves that keep
the dogs and me warm and cozy as housecats.

But storms do not allow much in ways to comfort,
not for long anyway. The wind on this mountain
can literally howl. Snowpiles can collapse trees
and ruin pole barns and wire fences. This is
when the shrinking wood pile starts to worry you,
and the electric bills skyrocket with the use of
space heaters and water tank defrosters. Hay is at
a premium, and always in demand.

It is a harsh month, but a necessary one. It's these
weeks that make winter real and make it stick.

4 February. *Twice Warmed*

If I have any leftovers from last night's dinner, I don't fuss with the stove in the kitchen. I just slide the crock of mac-n-cheese or leftover soup into the Bun Baker's lower oven. Soon it'll be warming up with the same energy source I knew so intimately back in June, when I was out there splitting that seasoned locust, sweat running out of my pores like out of a tap.

There's an old saying: "Wood warms you twice." Few things are truer. You burn up an inferno of calories doing the lumberjack work of felling, sawing, and carrying the logs home with a harnessed horse, then you heat up again while you sit comfortably by the stove. While I do prefer that second warming, I can't help but admit it's because of how much damn work went into earning it the first time.

Eventually I plan to move off the utility grid entirely and use solar panels to heat my water, with wood only for household heat. But these goals will happen in smaller financial steps. When it comes to home improvements, I do what I can, when I can — and farm improvements always come first.

8 February. *Scrappy*

Cold Antler Farm is a scrappy place. The fences sag,
everything needs improvements, and the layout is
suspect at best. But I love it. You can see the front
of the house in its moldy winter glory. The plastic
siding needs to be scrubbed and my little stepladder/
mop idea doesn't hold up to proper homeownership
standards.

So I call the boys at Common Sense Farm to come
up in the next few weeks with their ladders and
fancy truck, and in two hours this house will look
a lot smarter. There are about ten odd jobs that
involve heights I cannot reach (without crying),
so as a woman firmly grounded, I welcome their
help. Since we have a barter system between our two
farms, there's a good chance it will cost me some
livestock, such as spring meat birds, but that's fine
by me. Cold Antler will always be a scrappy place,
but I will do my best to make it less so.

Hmm. Can you embrace scrappy while trying to
fight it?

11 February. *Thaw*

A thaw came today. The real deal. Fifty degrees of
slush and sunshine. You can actually *see parts of my
driveway.*

I am outside tonight with Jazz and Annie. The
stream that runs through my property is roaring
and the sound is spring. I stand there and close
my eyes. Poetry, that rambling. Mixed in with that
gurgling percussion are the soft hoots of a Great
Horned Owl in the high trees.

I turn around to look at the house. It looks tired:
stains from the old furnace pipe on the front, leaves
and ice stuck to the sides. It looks like a panting
version of the proud white house that shone like a
lighthouse on the green mountain all summer. Yet
she is still there, and I realize that we have almost
made it through our first winter here, this pack
of four.

Gibson is on the steamer chest in the window,
sitting on his sheepskin and watching us. A light
is on upstairs in my office. It is on because I forgot
to shut it off after I watered the snap peas; but it
is also on because somehow through all the snow

plows, heating bills, and mortgage payments, I have managed to keep paying the electric bill. Same goes for the Internet, groceries, and gas for the truck.

Such modest accomplishments, but I feel like a domestic superhero. This place is making it. We're not eating lobster dinners or keeping the heat above 64 degrees — but we are making it. And who needs 64 degrees of luxury when it's a comfortable 34 degrees outside after dark!? Hell, I don't even have a jacket on.

But time to slow down, breathe deep, and come home to a banjo and three smiling dogs is all I need. Throw a thaw on top of that sundae and you've got yourself a farmer crouching to pounce on any defrosted soil. Soon it will hold potatoes, peas, lettuce, and more. Soon lambs will be running across the fence lines in little gangs of lost boys. And soon I'll be running again across the back roads of Washington County under a blazing summer sun. It just takes me a little time to recharge to see all that.

Tonight I see it. I might even get six hours of sleep tonight.

Take that, winter.

13 February. *Sleigh Ride!*

The thaw is over, and it ends with a snowstorm!
It has been such a mild winter, snow-wise, that
the blizzard comes as a shock. Patty invites me for
a bona-fide sleigh ride with Steele. Within twenty
minutes of receiving that phone call I am bundled
up and outside her barn ready to jump into the
Victorian cutter she bought at an auction. It is a
thing of beauty, black with plush red velvet seats.

Today as Patty and I harness Steele to his 1880s
sleigh, I feel for the first time that everything makes
sense. I know where to clip the breaking pull-backs
attached to the shafts. I remember to unhook the
girth before running my lines through the loops.
And the whole time I feel like a friend, not just a
student. Makes you feel lucky.

I am living in a Grandma Moses painting.

16 February. *Eggs & Carrots in My Pockets*

Cold Antler is both a modern farm and an old-fashioned one. What pays the bills are books and workshops and ad sales, but what fills my stomach and my soul is the land. I am always cash tight and never really know where the next month's mortgage payment will come from. But I feel rich for this piece of land that feeds me and the gifts, lessons, and life it offers me. Even on a frigid winter morning there is hay to haul and eggs to collect, chicken in the freezer to defrost, and a bin of homegrown taters in the pantry for a side.

The geese cackle and Gibson races around me as I carry buckets of fresh water to fill the heated, ice-free tanks for the livestock. I pull a small carrot from my back pocket and offer Jasper a snack. His brown eyes light up at the flash of orange. Cars on their way to work and school drive down the mountain road, and I wave to people I know. My own office is upstairs in the computer/tack room, where saddles and harnesses line the wall around my desk. I will go there after everyone's had breakfast, myself last of all.

I come inside with chattering hands to the warm stove-lit room melting with eastern daybreak and can nearly fall to the ground in tears. Eggs and carrots in my pocket — this is work and salary in one beautiful unit of understanding. I am home and my home feeds me. My life can be stressful and scary, but it makes sense.

I wish everyone could feel this and know.

25 February. *Eyes on the Prize*

Patty and I have become fast friends. As the weeks
of harness and driving lessons have continued,
I now associate winter Sunday afternoons with
backcountry roads, heavy horse collars, and Steele's
jingling bell. Patty has promised that come summer
we can take him all the way to Battenkill Creamery,
a local dairy and ice cream haven.

We've had a lot of conversations about horses and
my dream of owning one. And the more I talk about
horses and drive hers, the more I realize that I can
do this. Find a calm draft horse and hitch him up
for rides into town. And not just any draft horse.
I want my dream horse: a Fell pony, that horse from
the picture books I thumbed through at bookstores.

All I have to do is come up with the money, barn,
training, and time to own one. Yes, it's far from
realistic and even further from reasonable, but my
eye is now on the prize, my dreams foaming at the
mouth. You never know. Magic happens.

26 February. *Pig Harvest Tomorrow*

There are a lot of sighs on pig harvest day. It's intense and not at all enjoyable, but you accept those sighs. They are decisions exhaled. You own them, and you move on. Better sighs are just around the corner. Tomorrow will be a dark day in the life of my pig, but a bright one for this farmer.

I'm proud of this Pig. She grew fast, fat, and true. A trio of experienced traveling butchers will be arriving and taking care of everything from killing to skinning and halving.

I feel blessed to have these folks just a phone call away. For animals, most of the stress of slaughter isn't the agony of death but the transfer to the abattoir. Animals get confused and scared from such change, and may spend a day or two waiting in concrete stalls, cramped or fearful, while they wait for their ending. My pig will die in the same place where she has spent the last three months sleeping and eating. It will surprise the hell out of her.

There are worse ways to go. I won't pretend slaughter-day death isn't horrific; it is. But when done right it is as quick and humane as possible, and from the moment the gun is fired to when

the animal is gone from this life is literally two
minutes. It will be just a sliver of her time here at
Cold Antler. I don't like watching it, but I always
do. I feel it is my responsibility to be a part of the
whole process, from holding a piglet in a dog crate,
squealing in my arms, to the day its head lies on a
snowbank.

Even with all this, it's still a day I look forward to,
and I mean that without any harshness or disrespect.
Today is the day the work of raising this animal is
done and she will serve her purpose. I started the
day singing, and I will end it a bit more somberly —
but not without joy. The death of a pig is a cause for
celebration, feasts, and the promise of more piglets
soon. Pork helps keep this farm going strong.

27 February. *Red Snow*

This morning I wake up to thick snowflakes
coming down, snow-globe style. It is beautiful. I
stoke the woodstove, put on a pot of coffee, pull a
wool sweater over my braids, and head out to see to
morning chores. I think about the Tschorn family
hosting dogsled rides this morning in Bennington,
and a coworker's photography show I still have not
seen. But today I won't be riding in a dogsled or
looking at pictures on a café wall. I go into the barn
to have some words with Pig. Yes, I named my first
pig *Pig*. Not very original but true all the same.

The day is warmer than I am used to — already
in the high 20s, and the comfortable atmosphere
paired with the gentle snow seems to soften the work
ahead. Pig, not yet six months old, is just inside the
old red barn door. She is standing up, looking at
me curiously. She lets out a few gentle snorts, and I
realize how quiet she has been all along. She never
really makes any noise unless something in her
evening meal makes her snort with glee.

Despite once eating a hen alive when it flew into her pen, she was never vulgar. She didn't smell bad. She wasn't violent or jumpy, and never complained — though her world isn't one to cause much complaint. She lived a peaceful and comfortable life in this barn. Her nests of hay, pan of grain, and red water bucket served her well. Compared to most of the pork in this world, she has lived the life of royalty. Her little curled tail wags as I scratch her ears.

I look down at her eyes and say, "Thank you." I have never meant those words more than at this very moment. I walk out of the barn. The next time I go in will be with the butcher.

No. 150.

Winter.

A. G.

CHARLES GOUNOD.
(Arr. by Adam Geibel.)

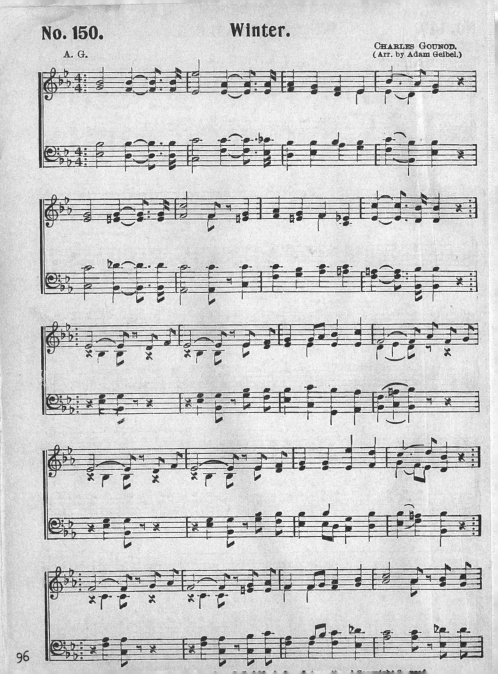

28 February. *Full Freezer*

Harvest Day was an event. Eleven people and six
dogs made up the work crew that descended on
Cold Antler. The mission: turn a living Yorkshire
Pig into food resting in the chest freezer. The pork
patrol was mostly limited to me, my good friend
Steve, and Vicki, the traveling butcher, and her
husband. But other friends of the farm stopped by to
help with various aspects of a winter animal harvest,
and their effort was greatly appreciated. Two
families arrived with their young children and did
everything from help pick out a butchering location

to climb on the roof to push off snow (necessary to keep the pork-cooking kitchen roof from leaking). It takes a village, darling. Welcome to the table.

This morning after my first hog harvest, I have to check the freezer to make sure the meat is still there. I'm not entirely sure it wasn't a dream. I want proof positive that yesterday's work party actually happened. But when I crack open the chest it is all there and accounted for: more than 120 pounds of roasts, ribs, hams, pork belly, sausage fixings, loins, and chops. I have never been in possession of this much meat before in my entire life. I am lousy with pork. I get a little dizzy and shut the lid.

Later this morning, fetching water from the well for the sheep, I walk past the giant maple tree. New snow has all but covered the blood. A farm exhales.

SPRING

I DO TOO MUCH

"You do too much." I am told this all the time. I have too many hobbies, too many obligations, and too many animals holding me down on this farm. Sometimes I believe this. Sometimes. If I just kept a few chickens and some raised beds with a couple or three sheep, life would be easier.

And I would be miserable.

I do what I do because it fills my mind, body, and spirit. I live in this frenzy of activity not as a victim but as a celebrant. Spring is on the way and now the party really gets started. I have chicks on order, seeds to gather, and then the real work of coop and garden. I will get a couple of new feeder pigs, and I'm preparing for a pile of lambs! Yes, much is happening. It is all wonderful.

Some days are overwhelming and scary, and those words "too much" haunt me like ghosts. They keep me up at night. But every morning I know what I am capable of, and what this farm stands for. What

feels like fear today is inspiration tomorrow and
nostalgia around the fireside in a season.

I'll figure out the mortgage, the freelance, the bills,
the manuscripts, and the workshops. I'll deliver
the kids and the lambs. I know bright spring is just
around the corner. Yet this in-between time makes
me jumpy and doubtful of myself. It's not what I
have taken on that scares me, it's that I'm not doing
enough. Not doing enough to make this farm work,
to make myself healthy, to make mistakes disappear.

You know what I think? I think wasted potential is
a lot scarier than feeling overwhelmed. There is no
monster greater than regret. I wouldn't wish it on
anyone.

Yes, I do too much. It's what I do.

march

Right now Cold Antler Farm is a
weird place. Outside it is cloudy and
mild, and the wind keeps picking up.
This lets little slivers of sunlight hit
the slush puddles just long enough to
lift my spirits before the wind covers
them up with clouds again. What
results is a bipolar weather pattern
we call early spring.

It is March.

4 March. *Mud and Maple Sugar*

It is not spring but change is in the air.
There are days with a bit more heat in the
sunlight. The lack of biting cold is noted.
The sugar maples know all about these
shifts and start to let that sap rise. Sugaring
Season is in full swing, and another
agrarian festival is at hand.

Sugaring in many ways defines an equally
eventful but disdained time of year: Mud
Season. As the sap rises, the frozen earth is
thawing. Late March and early April find
our dirt roads a mess of deep ditches, thick
brown slurry, a new vintage of potholes, and
nearly impassable obstacles. The towns do
their best to level them out and dump gravel,
but the warm days and frozen nights — while
great for sap — are a powerful force.

Best to stay home and enjoy another round
of pancakes.

5 March. *Unicorn Sighting*

Sitting at my computer, I postpone work by checking
Craigslist for my dream horse. I type in F-E-L-L
P-O-N-Y as usual, expecting to see the same old
cowpokes, backyard ponies, and pricey hunter-
jumpers I always see. And suddenly there he is:
a 15-year-old black gelding named Merlin, located
30 minutes south of my farm.

But the asking price is nearly eight grand. If I had
that kind of money, I'd use it to rebuild fences or
pay off my pickup truck. Staring at Merlin's photo,
I post the link on my Facebook page and jokingly
ask, "Anyone have eight grand I could borrow?!"
A reader soon suggests, "E-mail a counter offer.
The horse market isn't what it used to be."

Some sprocket in my heart clicks into place. I
e-mail Lisa, the owner, and pour my heart out. It's
darn close to begging, but I have nothing to lose.
I hit SEND with the anticipation of a junior high
school girl asking a boy to a dance.

6 March. *Real Magic*

Lisa has e-mailed me back. She loves Merlin, she
wrote. He was her dream pony, too, and the right
home is more important than anything. She said
to come see him and we can talk terms.

Suddenly the horse is no longer just a photo but a
distinct possibility. This is how real magic starts:
you make a wish, and out of nothing but hope,
horsehair appears between your fingers. You may
not own the horse, but to go from visual to tangible
is the difference between loving a song and learning
to play it on your own fiddle. It's more complicated.
It's more *everything*.

I saw him this afternoon and I fell in love. Hard and
fast and certain. Tonight I will pray, plan, and pray
some more. I still have no idea how to afford him,
or the costs of vet checks, boarding, and lessons,
or how to prepare for a working draft horse. At some
point, though, if I am the right woman for Merlin
and he the right horse for me, we'll become a team.

7 March. *Terms*

This morning I sent Lisa a heartfelt, honest letter, telling her what I can afford as a selling price and suggesting I try a three-month free trial. If the vet, farrier, trainers, and I all feel Merlin is right, I will buy him with a down payment on June 1 and make small monthly payments for two years until he is legally mine.

This would be a win-win situation: I would have three months of as much time with him as I could spare. Professional trainers at Riding Right Farm in Cambridge would evaluate him and train both him and me, and I would learn how we worked together. Vets would check his overall health. And in the end if I didn't want him, Lisa gets her horse back with three months of professional training and an updated vet record.

The magic happened. She agreed to all of my terms!

10 March. *Lambs Soon!*

We are fewer than 10 days from lambing here
at Cold Antler Farm. This shepherd's records
for Blackface sheep mark the earliest birth on
March 19. All my lambing books say to take
that date and remove three as the earliest birth
day. That means by next weekend there could
easily be two more sheep at this farm.

I need to remain extra vigilant and make
sure everyone has what she needs to get the
job done. All the ewes are eating well, with a
nice straw bedding in the shed, and for once
Mother Nature is starting to agree with this
whole lambing thing. Tomorrow might reach
above 50 degrees. Beats being born when
it's minus 8.

11 March. *What I Think About at 2 AM*

My job when the lambs come is to make sure they are healthy and upright. If a lamb is by her mother's side, drinking her milk, I can assume all went according to plan. But not every domestic creature has a pristine labor. Since we have engineered these animals through selective breeding for traits like wool or lamb chops, the trait of smooth delivery hasn't always survived. Many popular breeds of modern sheep thus need a midwife on hand, just in case something goes wrong.

Most of the time (especially with my hardy Highlander) nothing goes awry, but I always stay vigilant. All I can do from tonight onward is set my alarm for 2 AM — and then 4 AM — to check on things.

Regardless of how the birthing process goes, once the lambs are on the ground, there is work to be done in the first 48 hours of life. Each lamb's umbilical cord must be trimmed and dipped in iodine, and the lamb has to be tagged, docked (its tail removed), and looked over.

If I find a lamb in the shed or the snow without its mother, I'll just have to pray that it's still alive and will bring it indoors under a heat lamp on a blanket. I'll feed it some of the frozen colostrum I have on hand. Just in case of such events I have special tubes that the lambs swallow down to their stomachs, along with formula if their mothers don't produce milk.

Any day now. Any day now. . . .

13 March.
How to Never Feel Poor

Hundreds of tiny seeds are outside in
the little greenhouse: peas and lettuce,
kale and spinach. I grow heirlooms
because I feel as if I am growing secrets.
These plants are only for those willing
to seek them because you can't find them
in stores. My Amish snaps, my Rocky
Top lettuce, my Russian kales as purple as
Puff the magic dragon. They are sleeping
babes now, under warm comforters of
soil and sunlight. But in a few days there
will be a sea of green life.

It never gets old. You never feel poor.

16 March. *Old as Fire and Song*

There's a good reason shepherding holidays fall
in the order they do, and each one is just a part of
the annual cycle that is ewe and ram, fleece and
lamb. Care of the flock breaks up my year now the
way school and church holidays once did, and I
find myself as wound up and restless in the weeks
before Lambing Day as I was as a child waiting for
Santa. That day of lambing will bring something
magical, something traditional, something tangibly
wonderful and a part of me and endless people
before me.

What a thing! I feel as though a string of traditions
as old as fire and song are wafting back to me. I need
to learn all the particulars, but I have a lifetime to
do it. What more dare I ask for?

19 March. Due Date!

Today is the official due date for Martha. She's round,
and has a huge bag and a red rear end. It could
happen any minute now and, like the midwife I have
turned into, I am preparing for a midnight delivery.

They predict 16 degrees tonight, so I run a heat
lamp and extension cord up to the lambing jug
on the hill. That little brown shack is loaded
with hay, a water bucket, and a safe gate to keep
mother and young together. I bought a nutritional
supplement for the first 24 hours of life. I'm
ready, at least physically. I have no idea what will
happen emotionally once those little ones are
in my arms. I'll probably cry more than I have in
a long time.

This farm is a mess of mud and melting snow.
There are jars of honey glowing in the afternoon
light on the window sill and I know another
hive is on the way. The animals are all on clean
bedding, fed and watered. The eggs are collected
from outside and in the fridge. Production is
good. The lambing basket of gear and supplies
by the back door is like a hospital suitcase for a
mother-in-waiting. The dogs are asleep.

The chores are all done and now there is nothing but anticipation. *Sweet, writhing anticipation.* This farmhouse is humming with it. Any minute, hour, or day (even this minute as I type!) a ewe will start hunching with contractions and go into labor. I'm hoping I am able to be there to watch and assist (if necessary) as the first-ever Cold Antler lambs come into the light. If one arrives today, by morning it will be tagged and docked, given a booster and a head scratch. I will have completed the shepherd's year, and started a new one.

20 March. *Domestic Goddess*

The house is a mess. The kitchen is atrocious. No time to be a domestic goddess because right now the farm is a birthing center. I am head midwife, and a pair of twins (a ewe and a ram) are my entire world.

They were born healthy and strong this morning, and I moved all three into the small sheep shed on the hillside. This is my makeshift lambing jug, where newborns and new moms can get acquainted. There on a bed of clean straw, with a bucket of molasses-infused water for their mom, the little ones learn the work of a lamb — eating. It takes a few tries but both have gotten the trick of it and are devouring as much colostrum as their tight round newborn tummies can handle. Martha, the mother, laps up the sugar water, looking rightfully worn out.

Tomorrow I will band their tails and make proper notes, but for now I let them just be a family. It's a peaceful, blessed sight for my exhausted eyes.

Perhaps a little domestic divinity is here after all, and her name is Martha.

21 March. *Spring Equinox*

Summer is all about life in its prime. Fall is a
celebration. Winter is respite and repose, a time
to tuck in. But spring is a necessary and messy
flux. Folks all around me are excited about sunshine
and gardens and longer days, but I find spring
unsettling. It's like being excited about a new puppy
with no idea how much work and mess is ahead
before you have a good dog. And this time of year
is the furthest we get from the celebration of love,
light, and life that is October.

With that confessed, I always look forward to the
season's official start on the Equinox. The only
way out is through, and this banner day means the
promise of more light than dark. The day and night
are equal now, and daylight is the one growing
stronger.

31 March. *New Partner*

I have decided to make a goat part of
this farm, and I have chosen Bonita, a
three-year-old French Alpine doe who
will come to me from Common Sense
Farm. The farm is happy to sell her because
while Bonita is a great producer, she's also
larger and bossier than is preferable for
their needs. It's a match made in heaven
and I am grateful to get hold of a star milker,
at a discounted price, all because of her
attitude. I can tangle with a sassy goat. I
tangled with worse back when I worked in
a corporate office.

Before bringing her home, though, I will
need to learn to milk.

april

If I were a grower or a conventional farmer, and
vegetables were my main business, the calendar
would cut a wide berth around the crazy month
of June.

But Cold Antler is a small homestead, not a
conventional farm, and certainly not based on
vegetable production. My 6.5 acres of pasture,
woods, and forest center on the production and
life of animals, and it is not June that brings
me to my knees. It is April.

April can be a beast. The weather is unpredictable,
the tasks many, sleep rare, and everything that must
happen over the next six months depends on how
organized and prepared I am for the big A. Chicks

ordered months ago are en route, to be set into
brooders for egg and meat production. Sheep
need to be shorn — and by the end of the month
they will be bursting with more new lambs. My
garden must be planned out on paper (yes, even my
humble gardens need blueprints). Fences require
serious repair after a winter of fallen trees and
stretched wires; otherwise, easy escape hatches and
budding greens will tempt the sheep out of their
holding and into the road.

On any April day I may be running on three
hours of sleep. Placenta stains on my jeans,
I'll be wrangling sheep back into a pen while a
bottle-fed lamb bleats from one side of the
fence and his deadbeat mother tramples the first
shoots of mesclun mix on the other. Bedlam.

1 April. *Milking Lesson*

I am at Common Sense Farm, three miles down the
road. With Bonita locked and loaded in the metal
stand, my milking lessons begin. I am handed a
warm, wet rag dipped in lavender-castile soap to
clean and gently massage the teats before milking.
This "lets down" the milk into the teats. It is a quick
massage, followed with a drying towel. I use my
thumb and forefinger to make an okay sign around
the teat's base, pinch it off, and one finger at a time
close my fist.

Milk squirts into the pail!

Holy crow! Strike that. Holy goat, I'm doing it!
I keep at it and quickly learn how little pressure I
need. In the time it takes me to get $1/8$ of an inch
of milk in the pail, my partner at the other stand
is done and onto his next doe. So I give it some
moxie, and a little more gentle force, and the milk
comes out faster and thicker.

Okay, I'm getting the hang of it, I think. My arms aren't
used to the motion. I get tired quickly and have to
switch hands. As the pail fills up, I can feel Bonita's
udder emptying. When nothing more comes out
of the left teat, I focus on the right (which has plenty
of milk left).

In about 15 minutes I have finished the job. The
pail is foaming with a beautiful half-gallon of fresh
milk and I feel as though I just won a game of chess.
I have done something complicated and timeless,
something people have been doing for thousands of
years, a dance between hunger and style. And the
result is right there in front of me.

Checkmate.

Oh three wise men they came from the east,

To plough for wheat and rye,

And they made a vow, and a solemn vow,

John Barleycorn should die!

They laid him in three furrows deep,

Laid clods upon his head,

Then these three were rejoicing then —

John Barleycorn was dead.

— TRADITIONAL ENGLISH

7 April. *Old Songs*

My growing relationship to living seasonally is
in the doing. When you wait all winter to plant a
garden, finally hoe up sod when the ground thaws
in March, see seedlings burst forth in May, and
harvest sweet corn in August . . . well, all those old
songs and stories based on past agrarian culture
and religion seem more relevant than anything else.
So I have learned the traditional songs, like "John
Barleycorn." I play my fiddle with a bit of Celtic
flair, and as the old songs end long days of collecting
chicken eggs and planting zucchini, I feel timeless
and whole.

9 April. *Getting My Goat*

This morning Brother Zyrah from Common Sense Farm
delivers Bonita, who's wearing a collar and leash in a large
dog crate. Foolishly, I haven't anticipated the need for a
milking stanchion to secure her head while she is milked,
so I hold a grain bucket under her nose while Zyrah tends
to her other end. He has her milked in about five minutes
flat. I cannot believe his speed, three times as fast and
forceful as my beginner's hands. He carries the full and
foamy pail to the side deck for safekeeping.

As I go about morning chores, I rack my brain to figure
out a quick, inexpensive solution to the stanchion problem.
I have a moment of panic — *What the hell am I thinking, taking
on a goat without anything but a borrowed steel pail?* — that I let last
another 45 seconds; then I snap out of it and go into town.

At the hardware store I see a swivel flagpole holder, to be
wall-mounted with a pole stuck in at any angle you want —
basically a lever you tighten at will. Something clicks
in my head, and I buy it. Back home, I mount the holder at
goat-head height on the outside barn wall. Instead of a
flag, I stick a plunger dowel into it and tighten the screw
that will hold my "flag" in place. With a grain bucket
hanging in front, it will accomplish what the fancy metal
milking stands do at Common Sense Farm. Or so I hope.

10 April. *Stanchion*

It's milking time, I'm on my own, and I'm a little
nervous. I have no one to watch or help, and I'm
depending on an $8.99 aluminum flagpole holder
with a plunger handle attached. I roll up my sleeves,
wash my hands, grab the stainless-steel milk pail
and a plastic bucket of sweet grain, and walk over
to Bonita. Her bag is HUGE, and she seems happy
to see me, all bleats and head bobs.

I set her in place over my "stanchion," her body
against the wall, and clamp it comfortably shut,
using a piece of baling twine to keep her from
lifting her head out. She doesn't flinch, just goes
right to Chompsville. Setting the pail under
her teats, I go to town on those suckers. When I
realize I don't have to be delicate, I milk faster and
harder — and she just eats. In about 10 minutes
the work is done. I set the pail aside, hug her, and
help her out of her headlock. Mission accomplished!

Standing in my kitchen, I realize I need to strain
and bottle the milk. I don't have a regulation
strainer, so I substitute a small mesh metal strainer,
put a coffee filter in it, and slowly pour in the
foaming milk, noting that every goat hair and fleck
of hay stays out of the Pyrex bowl below it.

I'm actually doing this! Its final destination is a recycled bottle from Battenkill Creamery that I set on the kitchen counter, gawking at it for a while.

I just made milk happen. Me. After a few moments of quiet reverence, I put it in the fridge.

11 April. *Heirlooms*

Why heirloom lettuce seeds instead of my usual
six-packs of started Buttercrisp and Romaine from
local greenhouses? Well, this year I intend to save
the seeds, planting them again in the spring, and so
on and so forth into eternity.

Few folks realize that most of the vegetables grown
in America won't grow again from their own
seeds. They have been bred into a hybrid form
that produces just one generation of outstanding
product. So if you want a garden that can feed
you for more than one season, you need to plant
heirloom, open-pollinated seeds, saved by folks
who kept the old breeds of vegetables alive.

It feels good to be back in the soil again. I missed it
so much.

Is it just me or is it kinda creepy that most
vegetables from the grocery store can't be
replanted? I think potatoes with eyes might be
one of the few things we save from the grocery
store we can actually resurrect.

12 April. Chilling

I have started to enjoy Bonita's milk as a substitute
for everything I once used cow's milk for. In my
cereal, by the glass, in my coffee, her milk tastes no
different in flavor or texture than 2 percent cow's
milk. No "goaty" taste at all.

My now tried-and-true method: I ice the stainless-
steel sink first, half filling it with cold water and
ice cubes, and then I milk while Bonita eats her
grain and minerals. When milking is done, I set
the little flat-sided stainless steel milk pail in the
ice water. When it is chilled, I strain the milk into
glass containers and put them in the freezer for two
hours. It comes out slightly frozen, but I cannot
tell the difference between this chilled, fresh Alpine
milk and the stuff I've been drinking my whole life.

13 April. *Breaking Ground*

Leaning against the side of my truck, I let out a
very long, very tired sigh. In front of me on a slight
incline sits 32 square feet of future vegetables. I
have spent most of my Saturday constructing two
4-foot by 4-foot raised beds out of scrap lumber
I found on sale in the back of Home Depot. But
before the cordless drill meets any of the 2×4s, I
reenact the yearly passion play I've presented since
I lived in Idaho: breaking sod.

With my brand-new hoe I pull apart the earth and
discover black loam and earthworms underneath.
When about 10 inches of soil is loose and free of
roots and rocks, I fill a wheelbarrow with year-
old rabbit compost from the barn. It is covered in
decayed hay stained with Pig's blood. Shit and blood
are unpleasant things, but to a gardener they are
poetry. Left alone to think about their origins, they
decompose into a rich and beautiful potion that
literally creates new life. I mix in the unpleasant
with the raw earth and think about the rabbits, Pig,
and the months of story that go into a bed of lettuce.
What a thing this wooden frame will be.

I cover the earth and compost with 6 cubic feet of purchased organic black topsoil. I make five long mounded rows and plant the seeds a half inch or so below the dark earth. How odd to be engaging in this ancient practice with heirloom seeds ordered online! And what a great time to learn older country skills. Between the Internet and our force of will, we can learn or achieve just about anything we are stubborn enough to attempt.

15 April. *First Chèvre*

This weekend I use a kit from New England Cheesemaking to create chèvre from Bonita's milk. The kit came with cultures for curdling the milk, rennet, butter muslin, a recipe book, and four molds — all for 20 bucks. I thought making mozzarella was simple, but chèvre makes that cheese look as complicated as trigonometry.

I start after morning milking on Friday, and the cheese is ready to serve Saturday morning. Most of that time is draining and cheese curdlin' — only about five minutes is actual work. The hardest part is waiting!

In the morning I set the drained curds into a mold and let them drain even more, until they firm up into a freestanding mini-wheel of salted delight. I taste the cheese and close my eyes to savor it. It is nothing like any commercial goat cheese I ever ate. Light, fluffy, tangy — like slightly soured whole-milk cream and cream cheese whipped up together. There is no "goat" taste at all. I spread it onto an everything bagel and it is the perfect companion of light and fluffy cream to the seasoned and seeded bread. The hot bagel and the chilled chèvre do a dance in my mouth and I decide Bonita is here to stay.

Life is too short to live without a goat. And you can quote me on that, darling.

Here's how easy it is to make chèvre from raw goats milk:

1. In a large stainless-steel saucepan, heat 1 gallon of goat's milk to 86°F

2. Add 1 packet of chèvre culture & stir well into the warm milk.

3. Remove pan from heat and let it sit covered for 12 to 20 hours. The milk will set into happy thick curds you can slice with a knife.

4. Lay a piece of cheesecloth in a colander & set the colander over a large bowl. Pour the cheese mixture through the cheesecloth. Cover and let drain for 6 to 10 hours.

5. Add salt as desired to taste, not more than a few sprinkles. Place drained curds into containers and refrigerate for up to a week.

Produces 1-2 pounds of fresh soft cheese.

18 April. *Come She Will*

I fell in love hard with living by the
agricultural year; I have even learned to
love April, a month I always hated. But
that was before it turned into a world of
seeds in the earth and new lambs in my
arms. It's also looking a lot more appealing
weatherwise, when home heating isn't
my day's goal, and I can focus on things
like seedlings and violin strings.

April is mostly about new life. It's a
month-long holiday of rain showers and
seedlings, of mud and birthing fluids,
of newborn chicks and bunnies in my
grateful open palms.

20 April. *Earning One's Keep*

Inside the barn is a trio of rabbits I keep as a breeding set. Two does and a buck share a series of hutches, and a few times a year, the does produce litters of kits. I raise the rabbits for their meat, the most tasty, filling, and nutritious food a backyard can produce. Just three rabbits can keep you well fed — a doe can produce up to 70 pounds of meat (through raised offspring) a year. Two does means you're filling your freezer with more meat than a large whitetail buck can offer. Not bad work for three hutches in the barn!

So the rabbits have a very important and useful role. I can't say the same for the geese. My Toulouse goose and gander have been with me since I first moved to Vermont six years ago. I should have done some advance research because it turns out geese can live to be 40. Cyrus and Saro will be with this farm a long time. Occasionally they provide eggs, but mostly they are just my alarm system. Their work is to waddle about yelling at any stray cat, mailman, or guest who dares step onto their land.

I should have done some advance research because it turns out geese can live to be 40.

22 April. Ruminant Health Check

Bonita's udders have felt warmer than usual. Between that and the engorged teats, I worry about early signs of infection or mastitis. Since I am new to goats, I opt on the side of safety and call Common Sense Farm immediately.

Their farmer, 28-year old Yesheva, has become a walking goat encyclopedia and medic. She has been through it all, from diagnosing and curing Floppy Kid Syndrome (which as the name implies causes extreme weakness in newborn goats) to stillborn deliveries to extreme cases of mastitis.

Yesheva arrives armed with a thermometer, a strip cup, and her kind, gentle manner. She takes Bonita's temperature, which shows a low-grade fever of 104°F. Then she milks a few squirts into the strip cup to determine whether the milk has any strings or clots or odd coloration. It comes out normal, so Yesheva doesn't think it's mastitis as much as stress from the move — goats are not into change. But they are into sweet feed, so I can offer a little more to fatten her up a bit.

When she finishes her examination, Yesheva points out a scab on Bonita's right udder — exactly where the hands milk. She explains how to treat it, how important it is to keep Bonita clean and her bedding pristine, and to wait until after milking to give hay.

Apparently it takes about 30 minutes for the udder valves to shut, blocking off access to her milk and keeping out dirt and bacteria. A just-milked goat that slumps down on dirty bedding after a milking is asking for trouble. If she is standing up eating from a hay feeder after milking, she is more likely to be off the ground while the udders close up.

Becoming a farmer ushered me into the gospel of dirt, life, sex, and death.

23 April. *Vital Sign*

My day starts with taking the rectal temperature of
a goat.

Bonita is in her new stanchion, built by a handy
friend. She is chomping into a feeder tray of Dairy
Goat Ration, drizzled with molasses to distract her.
I hold the thermometer in my left hand and her
tail in my right and insert the thermometer. Bonita
keeps chewing, totally focused on her breakfast as
the device calibrates her body heat.

The thermometer buzzes, and I slide it out. 100.3°F.
Good news! We are back to normal.

This has been the only bump in the road so far, and
it wasn't a big deal at all. Onward with the milk pail,
friends!

Normal body temperature of a ewe: 101.5° to 103.5°F;
higher during hot weather

24 April. *High Holy Days*

Regardless of our different religions, all shepherds
celebrate the same holidays. There's Lambing,
Barn Building, Hogget's First Snow (I added that
one), and of course . . . Shearing Day. These are
the high holy days of the sheep calendar, shared
events understood by everyone in the Society of
Lamb and Wool. It doesn't matter where you live,
or what spices were stored in your family cupboard,
all wool shepherds are brethren when it comes to
spring rituals. Anyone who thinks ritual is dead in
a secular society doesn't have ruminants.

Tomorrow is Shearing Day, and we're ready to
testify.

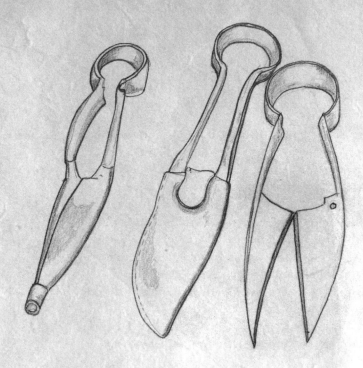

25 April. Shearing Day

Today the barber pays a visit to the flock at Cold
Antler. It is perfect weather for haircuts — mild
and sunny with the occasional crisp wind. For early
spring it sure feels like early fall. The shearers, Jim
and Liz, will be here around 3 in the afternoon. In
the small-farm sheep world, this is Midnight Mass.

When the white truck pulls in the driveway, I shake
hands and help carry their gear up to the sheep pen.
Within minutes the shearers are in their felt boots

```
HOGGET = a young sheep, usually between
6 and 15 months, that's never been shorn

WETHER = a male sheep, castrated when young
```

and set up with extension cords and blades loaded
in their giant shears.

Joseph is first to be flipped and shorn. For a first
timer, he is good. I can't believe how pitch black
the wool is under his brown locks. When all the
wool is off, Jim says I can bag it while he trims
the wether's hooves. My little hogget is a hogget
no more. Shaved, he looks like a little black doe,
a fluke, a sheep. I scratch his ears and tell him he
did well.

Sal and Maude are troopers, and after they, too,
are shorn, all three look like paper dolls with their
outfits torn off, awkward and naked and oddly
innocent. Sal, being Sal, comes up to me and leans
his big 200-pound frame into my thighs. My lion
has turned into . . . well, a hairless lion. Same thick
dope of a sheep but with less mane. He nips at my
shirt a little. I adore that ruddy animal.

Maude just stands on the hill and glares.

30 April. *Codependency*

The story of the goat is an old one. I've never had an animal whose exchange was so honest. This isn't about stealing eggs or jumping on a horse's back for a ride; this is a codependent and necessary relief. Bonita aches to be milked and I have a thirst that needs to be quenched. My labors milking her make her more comfortable, and filling the fridge with fresh raw milk makes me more comfortable. It is a fair trade, and one that will remain a part of my life for some time to come.

may

At some point the intensity of spring planting
and lambing sighs to a halt and the days grow warm
to the point where you take off your long-sleeved
shirt and hang it on a fencepost while you weed the
lettuce beds. And if you take the time to sit back
on your haunches and feel the sweat around your
temples, the sun on your back, and the softness of
the working soil below, you might, *just might*, realize
it is May. She made it. In the tussle of those chaotic
months of mud and birthing, there she was. And
when the hoe and the crook are by your front door
instead of always in your hand, you feel a sweep
of relief and gratitude just to soak in the rays, as
though you yourself are planted.

If you make it to May, you are.

1 May. Beltane

It's the first of May — to me a holiday and the first
real day of the growing season. They used to call
this Beltane, and some still do. I am among those
keeping that old holiday going strong. There's
no Maypole here, but I'll never forget reading
about how maidens hoping for love would wake up
on Beltane morning and wash their faces in the
grassy dew. Every May 1, I walk out to the grove of
forest behind the barn and wash my face with that
unbearably green grass's exhalation of dew. I believe
in magic, even the silly superstitious kind, and
regardless of its merit it feels nice to have an excuse
to stand out in a sunny grove and greet the morning
the way the herd of deer on the hillside does —
with dew on our nose tips and the love of the grass
in our heart.

I digress! It is the first of May, and a new litter
of rabbits has been born, out of Meg's Salad Doe
(the gray Chin's name is Salad) and my Silver Fox,
Gotcha. That makes three litters of rabbits! My
freezer won't have room for a fall pig with all these
chickens and rabbits. Not a bad problem to have!

2 May. *Gardener on Fire*

I have so much ahead of me in the garden, so much to plan and till and plant. So far only a bed of garlic, peas, and salad greens is popping up. But there will be more, and if I can get a rototiller over here, I will plant a proper farmhouse garden: corn, pumpkin, and potato patches.

These are the seasons to me. Green vegetables mean spring and summer. Corn means August and almost fall. Pumpkins mean pure fall. And potatoes mean winter. I want them all because on a bitter winter's day potato-onion soup tastes so much better when the potatoes and onions are your own.

Come May it is time to get serious about vegetables. This is when I plant the bulk of the garden. Either I buy started peppers and tomatoes from the greenhouse down the road or I transfer seedlings I started myself. This is when squash and eggplant make room beside the pumpkins and the butternuts. I start a new rotation of kale and salad seeds in the shadier beds to hope for a longer harvest of the dinner side basics. The garden isn't ordered, but it is alive in vibrant ways no flower bed ever dares to learn. I also make time for sunflowers by the front door, corn in a patch behind the house that gets the southern sun, and basil and mint spilling out of containers by the porch.

I am a gardener on fire, and the spark is that warmth I missed all through the messiness of spring.

It would be fun to grow soup as a community. All of us can plant some potatoes and onions — farm, suburb, or inner-city pots on fire escapes — and can harvest, store, and make soup together in December.

3 May. Barn Rising

Friends and I are starting to build the frame and
roof of a small pole barn for the horses. It'll be a
bigger and more welcoming space for Jasper and
perfect for when Merlin leaves the fancy stable and
comes home. To save money and time, I plan to
leave off the walls, my reasoning being my ponies
don't need an enclosure in the summer heat, just
the metal roof of scavenged tin we will rig in case of
downpours, hail, or debris. I'll add sides and more
protection as we head into autumn. For now, it's
four posts and a roof.

If it takes a village to raise a child, it takes a tribe
to raise a barn. Big projects like this I cannot do
alone. Besides being the most unskilled carpenter in
Washington County, I am scared of heights, power
saws, and most pointy objects meant for repairing
things. Hand me a bow and arrow or a rifle and it
makes sense. Hand me a circular saw or a 10-foot
ladder and I tuck my tail between my legs. So instead
of fighting against the tide I invite, bribe, beg, and
barter for the help of people who work wood.

9 May. *Marking a Month of Milking*

I have now milked Bonita, my large Alpine doe, more
than 60 times over the course of 30 days, producing
more than 45 gallons of fresh milk! I can no longer
imagine buying milk from the store. Just like eggs, veggies
(in summer), and most of my meat, milk has wandered
from the realm of things I once consumed to things I now
produce. This little dairy is chuggin' along.

The time I spend with Bonita has helped grow our bond in
a way you just don't get from sheep. It requires an attention
that's between meditation and conversation, never one or
the other. I let my mind wander a bit, until a back hoof starts
to wriggle or there's a loud fart that reminds me to respond
to the animal my head is pressed against as I empty those teats.

And I like what milking is doing for my body. My forearms
are the most toned they have ever been — Downward Dog's
got nothing on Descending Udder. It's made my fiddling
easier too since I am using my hand muscles so much more.
I feel stronger a month into goat ownership.

So it's neither delicate nor brash. It is just what it is:
imperfect practice toward perfection.

Just like farming.

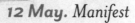

12 May. *Manifest*

Merlin and I are a true pair. Both of us a little old-fashioned, doughy, and out of shape. We look like shepherds, not Olympians, but a spring of teamwork will find us both in a better place, a transformed one. I know this because I have no doubt that my pony is magic — the old magic still running through a dreamer's heart and out past the forests in clear streams. Find a way to tap into it and anything is possible. And this horse, this Merlin, might be a magician of that old sort.

Magic is a fond desire coming true. Whether your Merlin is a vintage tractor, a mortgage, a new baby, *anything* — it is possible when you believe it is possible. Follow what you love with all you have to offer it, and the world makes a road for you.

I believe in magic because it is good for my soul. It gave me a horse I knew only in storybooks. It gave me a farm. It gave me readers. Someday it'll give me strong arms and a heartbeat to fall asleep against. I'm certain of that — as much as the black mane I brushed and kissed tonight. I'll keep the faith and I'll wait for the man. But I'm riding that pony tomorrow morning and I thank the ground we walk on for that gift.

15 May. *Hope Is a Garden*

Garden fever is setting in. It's mid-May, there's a gentle rain outside, and my spring-planted crops are coming up in spades. (The Arrowhead lettuce even *looks* like spades.) With peas, garlic, lettuce, potatoes, carrots, onions, rhubarb, and strawberries planted, this place is in the pink of health!

I hope an electric fence around the top sections, and the raised bed with some small ground fencing, will help with the rabbits and groundhogs — *hope* being the operative word. (I've learned the hard way that it doesn't take long for those dreams to die.)

But what is vegetable gardening if it isn't hope? A garden is telling the whole world, "Hey, I'm going to be around a while." It affirms life in proximity to your own home, and that sure is a beautiful thing. Even with the soil so far caked into the cracks of my hands that I can't wash it out, it is beautiful.

19 May. *Learning Curve*

The ritual of leading Merlin from his stall or paddock
into the lesson barn to be cross-tied, groomed, and
tacked up has gradually became second nature, a
routine we have memorized as if for a school recital.
First I brush him and remove all the dried mud from
his long mane, feathered feet, and body. I pick up his
hooves, inspecting them for wear and other concerns
such as cracking or overgrowth. I place his saddle
pad and saddle and tighten the girth around his belly.
The bridle slides gently into his mouth and over his
ears, his forelock cascading over the Celtic knotwork
on his brow band. I know how to check for correct fit
and adjust things where needed.

These are skills I did not initially possess, and it still
amazes me because of how much sense they make now.
The more you do it, the less you need to think about it.

25 May. Ram Holiday

Driving home with my new ram has been
wonderfully anticlimactic and *fantastically* boring.
For weeks I have been fretting about this particular
ride, certain it will be a disaster. I knew I'd have
an energetic Border Collie and a ram lamb inches
apart, crammed in the cab of a pickup truck for a
two-hour transport across three states. I prepared
for the worst. Gibson would be howling and clawing
at the crate, the ram screaming, truck swerving,
me praying as I slid down sketchy mountain roads.

I tried to prepare. I have a car-seat harness for
Gibson. I packed a first-aid kit. I planned to stop
often. I even started pricing stock trailers on
Craigslist.

All that worry has turned out to be for nothing. The
ride home is like driving through that Edward Hicks
painting, *The Peaceable Kingdom*. Gibson curls up in the
front seat, exhausted from his day of work, sleeping
like a babe of Eden. The ram lamb bleats here and
there but is generally resigned to his lot as cargo and
lies down. (I will soon find out he is "calmly" filling
the dog crate with liquid feces.) With my back to the
crate, my dog at my side, I am in a blissful state. My
truck is chugging through the Green Mountains like
a champ, and I am almost home.

The lamb I am driving home is next season's sire. He is a beautiful boy, a young Blackface ram. He's the breed I chose from all others to feed and clothe me. New blood, new lambs, new hope, and all of it tangled next to my chest when I carry him in my arms into the truck. Two hearts separated by wool, skin, cloth, and blood.

Picking up the spring lamb that will in time become the fall ram is new for this particular farm, but instantly ritualistic. It is one of those things you do as a new farmer and immediately understand you're taking part in the first of endless annual occurrences just like it. You are nostalgic in the present moment (which might be the closest to enlightenment this girl will ever get). My first Shearing Day was the same, along with my first apple cider pressing, lambing season, and that first spring hatchery order years ago in Idaho. They are holidays — holy days — you see.

Holy is the proper word, too.

21 May. *Lessons with Merlin*

My lessons with Merlin at Riding Right Farm
are going well. I'm far from an adept equestrian
but the calm words from Andrea and Hollie, my
instructors, are setting me up for success with
my new partner. Merlin has done all of these moves
before, he knows his walk from his trot and canter.
I am still nervous to move too fast or venture beyond
the indoor and outdoor arenas. There are trails
we can take together right beyond the white fences
but I still feel like a teenager who stole the keys to
her parents' car. It's exciting, but I'm not ready for
the open road.

I feel safe only when going out with someone else,
someone with more experience. If an instructor
or Patty wants to invite me for a ride I am so game
that it's hard to keep still in the saddle. But going
out alone is a Brave New World and I'm not ready
for it yet. I'll get there, though. I'm certain of it.

25 May. *Loading*

Merlin refuses to load into the trailer. Today
Patty's friend Milt comes to teach him how to
enter and exit the trailer calmly. It takes three
people, a rope around his rump, and a bucket
of grain, but we do it. By the end of the lesson,
Merlin and I walk onto that trailer without
fuss, just a loose lead rope in my hand and a big
smile on my face.

My goal is that Merlin become a second vehicle,
another way to get across the landscape. I want
a horse I can take to a neighbor's for a visit,
or hitch a cart to take into town. Patty still picks
us up for our weekly trail rides and field trips,
and Merlin and I work through our trials and
tribulations one at a time.

SUMMER

COMMITTING

I was not a religious person, though I respect
and appreciate what religion is. It's both a way to
live and something to live up to. Its observances
and festivals — if celebrated earnestly — make us
understand the world better and our place in it.
When I was a child, my holidays were full of magic
and great import. As I grew older, holidays became
merely Hallmark events. Soon, religion and I
parted ways. We still met up for coffee on occasion
but in platonic conversation. No commitments
from either side.

But farming is changing this. My life is entirely
about commitment now, and I find myself praying
more than ever before, mostly out of sheer gratitude
for my land and the air in my lungs. My prayers
aren't to anyone in particular, but they are constant
and honest. I make time to meditate now and read
through the sutras that make my head sing with
good things. The Diamond Sutra contains text
I want engraved on my tombstone if I get a say in
such things:

```
Thus shall you think of all this
fleeting world:

A star at dawn, a bubble in a stream;

A flash of lightning in a summer cloud,

A flickering lamp, a phantom,
and a dream.
```
                              — THE DIAMOND SUTRA

So my farm is bringing out a more spiritual side
of me, and it's varied and happy, and it makes its
own holidays.

We once knew how to eat in season, string up a bean
vine, dress a Thanksgiving turkey. Our children
were not scared of dead pigs but clapped their hands
under the hanging hogs, because they liked bacon
and because they weren't shielded from the whole
story. I want to go home to that mindset.

All this time I thought I was becoming a farmer,
but instead, I find the farm is becoming me.

june

We all know about the honeymoon that is June. Summer heat comes with long hours of daylight, and the farm sinks into a lull only a shepherd with a garden can appreciate. My heavy months are April and May, but come this sweet month my hardest work is for other farmers. While Cold Antler needs just milking, weeding, and the usual chores, other neighbors need hands at the hay wagons or rows of onions weeded out. I am happy to start the day in the barn, welcome the morning in the river with my fly rod, and then return home for a nap in the hammock and a light lunch before I am sent into the fields again.

This is June. Summer as it is meant to be.

1 June. *Waiting for Fireflies*

I listen to people around here who have planted
potatoes for more years than I have been alive.
While conventional wisdom says to get your spuds
in the ground early in the season, I wait. I wait
all the way through spring until we are flirting with
humidity and fireflies, and then my potatoes meet
earth. That is my rule. The reason is that both wet
springs and potato beetles cause havoc in earlier
settings. So I wait.

No tots in the mud until I see a lightning bug.

3 June. Swaying

This is a place where pure summer heat is a luxury
and a rarity. With a growing season only slightly
north of a hundred days, most of the year is spent
climbing into or recovering from winter. But when
the wheel of the year turns into June, there are no
thoughts of firewood and snow. There is just the
drunk and lovely waltz of fireflies and banjo frails,
accompanied by an occasional bass line of thunder.

I lose hours in my hammock, gliding above the
good earth, hugged by gentle ropes. Storms slide in
and out over the mountain, and when they leave,
everything is even more humid than it was. I find
peace in moments like this, a few inches above the
ground. I accept the humidity as a lover and lean
back into his arms.

As darkness falls, the farm turns shades of dark
green and brown. The only light is the flickering
of thousands of fireflies and the soft glow from the
light in the chicken coop nearby. From under my
King Maple I pluck my five-string banjo through a
few waltzes I taught myself from books. "Down in
the Willow Gardens" is my favorite. Played slow and

meditatively, it becomes part of the night. The music pours out of me like storm rain.

Fireflies, my backup angels, circle around the canopy of the giant maple that someone, many years ago, allowed to thrive alone in the front yard in the sun by a shallow well. It wants for nothing, that grand tree. It is the symbol of strength and acceptance on this farm that knows the cold better than this hot grace.

I dreamed of days like this when winter days stretched below zero and I could barely pull myself out of a hot shower. The memory of hardship makes the present more delicious. So I savor the night, and the swaying.

7 June. *Houseful of Honey*

My hive, a healthy work commune just
outside my kitchen window, is bustling, so I
am putting on a third-story addition. A few
nights ago I built the wax frames on my living
room floor, and today I am setting the new
shallow hive body on top of the two that are
already full of comb.

Since I am already out there working the
bees, I decide to harvest a wee bit of honey to
kick off the summer. I bring out my 5-gallon
brewing kettle, a knife, and a handful of
sheep's wool. After a proper smoking, I pry
open the inner lid to the hive (fused with
comb to the top hive body) and get to those
beautiful frames. The bees do not bother
you if you remain calm, and the smoke keeps
them a bit disoriented.

I pull out just two frames from the center
of the box, and with the wool I gently brush
the bees back into the hive. I re-set both of
my frames, then, like a fat and happy bear,
I waddle back to my house with my bounty.

With a large serving spoon I scrape the entire frame
into a metal colander inside a stainless steel bowl.
The honey takes about an hour to drain, and then
I do a second straining through cheesecloth.

I pour the honey into 8-ounce plastic bears where
they will wait in my cupboard until teas, fresh
baked breads, ice creams, and batters call them
into service.

9 June. *Balking*

With Merlin finally trained to load into the trailer,
it is my turn to learn something new. So I get into
a Western saddle for the first time for a trail ride,
and I balk at it the same way Merlin did at the metal
trailer. This Western saddle feels like a couch sitting
on the back of a rhino wearing a wig.

Feeling my lack of confidence, Merlin backs up and
circles. I jump off. Patty demands that I get back on
and not give up. I fight back tears as I climb back
into the saddle, and Patty works with him on a lunge
rope as I sit there, holding the reins like a child on
a pony ride. I've gone from a confident woman in
English tack to a scared faux cowgirl. Patty assures
me I will come around to the Western tack for trail
use, but I balk. I both hate her and love her for
making me get through today without quitting.

My horse is better at change than I am.

10 June. *Married to a Goat*

Every 12 hours that udder of Bonita's needs to be emptied or it will be painful, possibly get infected, and then dry out, and I'll be out of milk (and luck) until next spring when she kids. No more beautiful glass bottles of fresh milk (marked GOAT MILK) in the fridge. No more chèvre ready to spread over homemade breads and bagels. No more visions of milk soaps curing in the closet. Her gifts are mine for the taking, but my end of the deal is a twice-daily date with her teats. No exceptions.

So I am married to a goat. Every 12 hours my right cheek is pressed against her side as I milk and talk to her. She munches on her grain ration and sweet feed, and I relieve the pressure she feels. And you know what? She relieves mine. It is hard to be stressed out when milking any animal. I can't check my smartphone or worry about bills. I just milk.

15 June. *Merlin's Homecoming*

It took a pile of friends, hoarded money, scrounged metal and lumber, and a stretch of hot days, but the horse pasture is ready, right in time for Merlin's arrival.

Everyone agrees he is a gem, and we even entered a dressage show and won a third-place ribbon. It was the crowning moment of our three-month trial period and perhaps what convinced me in my heart to keep him forever. Now a prizewinner and a partner, Merlin is moving home to Cold Antler Farm. I make the down payment and am on my way to owning a unicorn.

Patty comes by with her trailer early this morning. Together we drive Merlin, tack, and gear back to Cold Antler, and I lead my new mount to his grassy pastures and his flatmate, Jasper.

Once in the paddock the horses run together, stopping to press their foreheads every now and then. They seem happy and I am beside myself. Not too long ago any horse was a pipedream, and now I watch the pair of them roll in the dust and play tag.

21 June. *Longest Day*

It is 8 PM on the longest day of the year, practically
bedtime around this farm where days start before
5 AM and charge hard all day long. My body is tired
but my mind is reeling; I am happy to stay up until
sunset and light a bonfire. Friends will be here
soon with covered dishes, folding chairs, guitars,
and banjos. We will eat, drink, and be merry under
that long sunset and say goodbye to the peak of the
daylight.

From here on out, every day a little sunlight fades
away. I am okay with that, thrilled really. All year
I am praying for October, and this is just another
grateful check on the year's to-do list. I lift a beer
to my lips and nod to the setting sun. He did good.

People who farm
have a different way of
understanding time,
one based on sunlight and seasons,
ebbing and flowing in activity
like river water.

22 June. *Bringing In the Bales*

Today I am part of a small team loading first-cutting hay into the hayloft of Patty and Mark's 1800s barn. We spend this whole afternoon stacking 300-plus 60-pound bales — which Patty prefers for her two heavy horses — onto a rickety elevator and up to the loft. Others are in the loft creating a Jenga mountain of 50-pound rectangles of dry compact grass. It is, at best, an anxious construction.

When finished, we walk up the hill to the stone porch beside Patty's beautiful white farmhouse. There we sit and drink cold beer and eat a gratefully laid spread and talk about the hay, the work, our farms, and our plans.

I discovered the hard way that hay is sharp enough to cut, and if you are foolish enough to wear a T-shirt (and shorts), your skin will be riddled with casualties from the chaff. So I now know to toil in long pants, long sleeves, and leather gloves for the rest of this haying season.

My long scythe whispered
and left the hay to make.

— FROM "MOWING" BY ROBERT FROST

27 June. *Joining Up*

Patty has convinced me to join the Washington County Draft Animal Association (WCDAA). She assures me that plenty of ponies, oxen, mules, and even the occasional donkey are among their ranks. I bite. If 11-hand ponies the size of Jasper are welcome, then a thousand pounds of British black thunder can surely show up.

I find that people interested in horse power are a friendly and like-minded ilk. And if the phrase "horse people" conjures up images of snotty upper-class overachievers, this is not that stereotype. The WCDAA consists of regular folks who just happen to love horses and traditional modes of transportation. Not everyone in it has a 401(k) or even a full set of teeth. Living with and loving working animals is the only qualification for membership (besides the $30 annual dues).

july

I like to think I spend most of July in the river,
even though that can't possibly be true. As the lone
operator of my little empire, I'm on my patch of
land 24/7 — until now. Come this hot and long
month I find myself taking off for the swimming
holes and fishing spots whenever I can steal a
moment. I meet the sunrise by casting dry flies
to trout in the Battenkill, and I meet the sunset
by diving deep to soothe a body tired from a day
of work, riding, writing, and farming. To feel
cold river water rush past cuts and bruises, swipe
off sweat and anxiety, is a baptism and a party
every time.

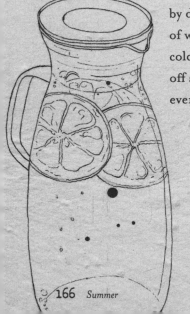

1 July. 100-Proof

If my summer existed as an arc — starting at late spring and ending with early fall — July would be the glorious peak of the curve. It's pure 100-proof summer. The hottest days of the year sneak up on us here in the Upper Hudson Valley like a jungle cat. One evening you'll be falling asleep on a cool 50-degree breeze, serenaded by tree frogs, and the next morning you're scrambling for cover under a shady maple to avoid passing out from the humidity.

Forget about the flirtatious lazy heat of June, with her fireflies and thunderstorms. You sweat a new kind of sweat, the type that literally pours out of your skin. You swelter and curse under the spell of the Evil Day Moon but by nightfall forget about all the swagger and gruff, get a cool outdoor shower, and meet up with friends in town.

It's in the evening that July sings to us, at dusk in particular. When everyone's work for the day is over and people are in the mood to greet and eat, it feels amazing to step into a cool mint-soap shower and then throw on a cotton sundress and cowboy boots. By the time the stars come out (and here there are *stars!*) you feel like a college student on spring break again.

That is, if college girls enjoyed a few cocktails before waking up at 4:45 to milk a goat.

6 July. Summer School

July mornings are now spent developing a new ritual with Merlin. Through lessons and the school of hard knocks, we have become a team. Every day, when the commuters on my mountain road have cleared out and traffic consists of just a few delivery trucks and random pickups, I saddle up my mount.

Merlin and I are not attached to any particular tradition or school. Even his tack is a mishmash of Western and English gear. Some days I slide a featherweight English endurance saddle over his wide back and braid his mane under an English bridle. I'll put on breeches and half chaps and a velvet riding helmet, and we'll trot happily down the road to break a morning sweat and give my thighs a workout.

Other days I get out a Navajo blanket and barrel-racing saddle I picked up at the local Poultry Swap and a latigo headstall with split reins. No helmet on these days, commonsense be damned. I wear a straw cowgirl hat, and instead of tying it under my chin, I braid the two cords into my pigtails adorned with chicken and goose feathers, feeling like a new tribe of woman. I put on my favorite Wranglers over dusty old brown paddock boots, and we hit my neighbor Tucker's fields and forest paths, getting lost among the old stone walls and wild raspberries.

10 July. *Staring Down Winter*

I'm a small farmer. My close friends are small farmers.
We have a variety of personalities, occupations, talents,
and vices, but the things that unite us are without
debate. Things like gardens, wood stoves, and livestock
are realities that have their own rules and regimens
governing care and upkeep. And even though it is high
summer — and no matter how much we love our fireflies
and barbecues — we never forget that winter is on its
way. You can tube down the Battenkill with a six-pack,
but you can't hide.

If you have a lot of money set aside, you can prepare
for winter in one hard weekend in September. You

order a few cords of wood and a couple hundred bales of hay and hire people to stack it for you. You call your mechanic and have snow tires ordered and mounted, arrange for a local plow to clean your driveway whenever it snows more than 4 inches, and order all the polar fleece techno-outdoor clothes your credit card can handle.

You can call the oil company, the chimney sweep, and anyone else you need for winter here in the North Country. With enough girth to that ol' pocketbook, you can have the closest thing to an endless summer that anyone could, up here. It's an illusion, of course: winter always shows up. But if you have the bank to buy that peace of mind at a drop of a hat, July becomes a lot more enjoyable.

I do not have a lot of money (set aside or otherwise), so my winter preparation starts in summer. It's easier for me to come up with the extra dough, little by little, for a cord of wood and 25 bales of hay a month, slowly stacking them, than to drop several thousand dollars in a weekend.

So come July 4th I am thinking about firewood. By the end of July I am obsessed with hay filling up the barn and side porch. Come August my chimneys will be swept and hay will be in. And by September I have cordwood stacked. That's my order and I stick to it; it's my winter prep canon.

25 July. Flight

On my mountain, on the back of my powerful black
horse, I am on top of the world. I ask him to canter
now, no longer afraid or timid. I know this animal
the way I know my friends and family. His mane
flaps like prayer flags as he lopes across the fields,
maybe jumping in the air to kick a little when I ask
him to. I hold on as we fly across the green grass.
Above us, herons stretch their wings and eye the
culverts in the brook for trout.

Merlin and I are two feral heartbeats, thriving far
from my home office filled with its deadlines and
bills. Out here are flesh and leather, horse sweat and
sunlight, the sound of heavy hooves on sweet grass
and the smells rising around us. No one told me
that a thousand pounds of animal walking through
wild thyme could send your nostrils an instant
sensory memory, like a Polaroid photograph.

As my handsome boy takes each step, I memorize the
moment like a Zen monk slowly chewing his temple
dinner. These are also Days of Grace. Any day that
you can't imagine being anyone else, anywhere else,
is holy.

30 July. *Swimmin' Hole*

In July the Battenkill River gets to its best swimming
temperatures of the year. It has become my happy practice to
work all day and then clean and refresh myself with a jump
in the river.

Near my farm is a swimming site called Pook's, famous to locals
here in Jackson. The farmer knew his corn met the river at
the perfect swimming spot, complete with big shade trees and
beachy sand, so instead of putting up No Trespassing signs, he
made a parking lot. He posted some information about safety
and picking up garbage, and not stealing corn, but that was the
extent of his rules.

The water rushes over your whole body, removing sweat and
grime. If you look down at your feet, you can see fish swim past.

This river is a lot of things to a lot of people: an oasis and a
relief, a trout stream, a place to pull irrigation for their hogs'
drinking water. Whatever it is, it is useful. Around here that
is a cherished virtue.

august

August is the end of my summer. Autumn is on his way, usually more in hint and spirit than in actual weather. Nothing is shouting *Fall!* into my ear quite yet. Instead I notice whispers vibrating off the squash vines as if from a string on a child's tin-can telephone. Walk outside for morning chores and take in the heat, but feel the change: there's less humidity, and the first leaves are already fading and hanging differently.

The garden and the forest itself are heavy with the peak of their growth, like a bowl of milk about to tip over and spoil. Senescence is just around the corner and the breakneck pace of growth seems now to apply only to weeds, which at this point I have let devour much of the garden without fuss.

1 August. *First Bread*

In medieval Britain they called the first few days of August *Lammas*, which translates from the Old Saxon as "Loaf Mass." Since the summer wheat was harvested before other fall crops, this was the official start of autumn in European agricultural societies. It's understandable on many levels, since people were not only bringing in the first big crop of the year but also grinding it and storing it in casks and sacks to feed their community through the dark months ahead. The wheat harvest became the first of several insurance policies for winter.

Although Lammas was a day of work, it was also a holy day. People baked loaves of bread from the first grains of the year and carried them into their churches to be blessed. They didn't eat the sacred Lammas bread but ripped the loaves into quarters and placed them in the four corners of their barns. There they sat as happy little sacrificial offerings. A tangible symbol of hope for a safe winter and good luck.

Lammas was a Christian celebration that stemmed from older Pagan ones. (Compared to farming, Christianity is a pretty new idea.) The Celts of

Pagan Europe held their harvest festival on the first, in honor of their Sun God Lugh, and they called the day *Lughnasadh* (pronounced "*loo*nasa"). It was a celebration of First Fruits as well as grains, since apples and grapes played a big part in that ancient story, too. The Vikings celebrated *Thingtide* earlier in the week. But everyone was focused on the same great burst of energy: grain harvest.

August is such a tease with her shorter days and cooler nights. I wake up expecting to need sweat-shirts and wool socks. Lammas and Lughnasadh announce the end of summer and the beginning of the great mystery of autumn.

3 August. *Second Cut*

Hay is baled and stored from June to August, and sometimes even later in the season if the sun and rain offer the right conditions. Here, first cutting is always taken in June and the hay is usually coarser, not as rich in nutrition as the later cuttings. Second cutting is everyone's choice for feeding hay, especially for horses. That's what I try to store up for my stock here at Cold Antler.

First cutting is cheaper but it's always less green, more like straw. Second Cut can cost up to two dollars more a bale, which adds up when you buy it by the truckload, but good hay is an investment in health. It's like beans and rice versus a pile of potato chips. The sheep will eat the potato-chip hay but they won't get what their bodies need.

So I suck it up and pay more for Second Cut and never regret it. Sometimes when I fork that beautiful, dry, green grass onto the snow in front of my sheep, it looks good enough to pour some balsamic dressing on and dive in with them.

HAY is green and used to feed animals.

STRAW is yellow and brown and used for bedding or compost.

12 August. *Helping Hands*

If a small farmer is asked to help with haying, she
shows up. Friends and friends of friends call and ask
if I'm free the following afternoon. No need to say
why. If there's been a stretch of dry days and you can
smell the cut fields as you drive into town, you know
it will be about haying.

You help because you can, and when you need
something later on, your help is remembered,
considered, and appreciated. In late winter, when
I'm down to my last 20 bales and the grass is a long
way from green, I know I can call any of the farmers
I helped hay and soon load up my truck with what
they have to spare for a fair price. That availability
is a gift in the ice-covered times of the year. Since
my barns can't handle the 400-plus bales I'll need
to get my animals through the winter, when I need
to reload my stash, I know there are larger barns
on larger farms that can help this little freehold out.
Having that absolute certainty is worth a few hot
days of toil.

Bringing in the sheaves, bringing in the sheaves

We will come rejoicing, bringing in the sheaves.

— FROM "BRINGING IN THE SHEAVES"
NATIONAL GRANGE SONGBOOK, 1929

15 August. *Spuds*

I love potatoes. I love the little white flowers that
bubble and pop out of the green leaves like the
frumpy girl in high school putting on her prom
dress. I love digging the tubers out of the ground
a few months later. I love holding their lumpy,
starchy meat in my dirty palms. I love the whole
reincarnation of the thing — the way a sack of last
year's uneaten spuds can grow warty new roots,
singing out to be sliced and placed back into the
soil. As long as your chunk of tater has an eye, it
might well sprout into a whole new plant. Which
means that little chunk can produce, instead
of a few hash-browns, 5 more pounds of glorious
potatoes. Not a bad second life.

Every year I plant a patch of seed potatoes, and my
goal is to plant more every time I do. This past
June, I planted my largest raised-bed garden with
33 seed spuds, and as August hails high, they are
leafy and thriving, about to flower.

18 August. *A Square-Bale Woman*

Few people in my small circle of local homesteaders
have a tractor. Being the broke traditionalists that
we are, this means our hay comes in smaller, square
bales instead of those giant 400- to 600-pound
round bales you see out in fields, covered in white
plastic or curing in the sun. The smaller size means
that a person can carry them where they need to be
without a gas tank. The giant round bales must be
moved on long spears attached to big tractors, and
my farm doesn't have any engine-powered obelisks.
And I'm located on a steep mountainside where
I would certainly flip a tractor within minutes of
starting it up.

So I'm a square-bale woman. I pick up bales in the
truck, unload them by hand, and move them around
the farm with carts and sleighs. And that's not horse
equipment, but handcarts and the plastic sleighs
you find in drugstores. I tie ropes to them and pull

If a small farmer is asked to help with haying,
she shows up.

or push the bales across the snow or fields. It is a
workout, and I like knowing after a morning of
chores that I have broken a sweat and made breakfast
for every critter in my keep. That never ceases to
restore my spirit and make me appreciate simple
things like hot-water showers and steaming mugs
of coffee.

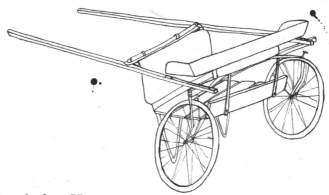

20 August. *Cart before Horse*

Patty and I are refurbishing an old farm cart that
she found at a local farm auction for five dollars.
A little paint, some elbow grease, a new set of tires
and rims, and I have myself one smart-looking
horse cart.

I found a Haflinger-sized harness on eBay for a
hundred dollars and spent another fifty having it
fixed up and soaked in oil by an Amish harness
maker from just north of the Adirondacks. Patty
sold me Steele's old collar, and William Beachy,
the harness maker, sold me some used tack to
finish the outfit, including a pair of 17-foot-long
biothane lines.

So for the unbeatable price of $285 I have a cart, a
harness, a driving bridle, and lines. Considering that
a new harness could easily cost $800-plus and a new
cart anywhere from $1,000 to $5,000, this is a steal.

25 August. *Turning*

Haying is my favorite work of the summer. All winter I will go into my little barn with its fading red paint and pull out bales, one or two at a time, to feed my animals. They depend on it. I depend on it. The hard work of summer may seem grueling, but it is the first strenuous step toward winter contentment. On a snowy morning, to feed your flock the very bales that made you bleed and pour sweat, months before, brings quite the moment of reflection. Then those calm summer afternoons become memories just as winter scenes were as I swayed in the hammock. That's how you know the wheel is turning and you're turning with it.

Stay on the farm, boys, stay on the farm

Though profits come in rather slow.

Stay on the farm, boys, stay on the farm —

Don't be in a hurry to go.

— FROM "STAY ON THE FARM"
NATIONAL GRANGE SONGBOOK, 1929

AUTUMN

HARVEST HOME

My farm is almost ready for winter, and I am anxious to get her completed so I can just lean back into the firelight, sip some warm cider, and enjoy the wood smoke and stories. Chimneys have been inspected and cleaned, hay fills the barn to capacity, and firewood has been ordered. But there is still feed to buy, wood to stack, and peace of mind to achieve as quickly as possible.

It isn't easy; in fact most of my time is spent worried and driven, getting from one bill to the next. Sometimes the emotional toll wakes me up at three in the morning, and on an otherwise beautiful moonlit September night, I am convinced that I won't be able to meet the expenses.

But then sunlight floods the farm, the coffeepot starts to bubble on the woodstove, and something inside me shifts gears. No matter what haunted me the night before, at first light the dogs need to go outside. The horses are already whinnying for their morning hay. The sheep see me stir and run down to the gate, their *baas* and bleats joining the heckling by the horses. The roosters crow, the chickens strut and coo, Bonita stands up on her metal fence, rolling her head and crying for grain, the pig in the barn snorts, and while the rabbits are quiet in their cages, I know they need their water bottles refilled and they're waiting for morning pellets. It is a circus and a symphony, and it does not allow self-pity or concern about anything that isn't happening *right now* to make 50 animals content.

With Gibson at my side, my day is new and my work is my mantra. I carry hay and feed bags. I dump buckets of clean well water into troughs. Within 15 minutes the cacophony of need is quenched, and chewing and clucks are the only sounds. Peace is restored through focus and action, my fear is stilled, and everything falls into place. It just requires a head down with ears back, facing into the wind, like a fox crossing a blustery hillside. You feel exposed, scared, but you go on because you must.

september

September 1st is like a gunshot start to a two-month-long harvest festival. The roadside stands fill up with pumpkins and mums, and wreaths of fall leaves and stars cover red farmhouse doors.

My own farm celebrates in subtler ways. Baskets of freshly dug potatoes wait in the kitchen, ready to be stored for the winter ahead. The garden is spitting out the season's last tomatoes and squash, and in the next few weeks the soil will be turned over and prepared to rest.

1 September. *Overripes*

Exactly 1.8 miles from my front door, the Stannard Farm stand is like having a farm-fresh mini-mart a bike ride away. They sell the farm's own meat, eggs, and vegetables along with local cheeses and milk, all priced to move for locals.

Today a 20-pound crate of slightly overripe/bruised tomatoes is $5. Jackpot.

Ella behind the counter says even when the fruit is slightly past its prime it's perfectly fine for sauces and canning. I tell her to hell with canning on the fly, I'll make a pot of sauce and freeze it tonight. That homemade sauce, defrosted and poured thickly over pasta, will taste just as amazing as canned when the first snow falls.

I bet if you ask around your local farm stands and growers, they'll sell you their overripes for a song as well. Worth a phone call, anyway. Think ahead a little and savor that sauce in advance. That's what I say.

2 September. *Stubborn*

When I leave the farm astride Merlin, the sun is starting to yawn and the world is bathed in yellow light. My black horse's feathered feet trot along the winding road and I close my eyes and take in a deep breath. These are the last deep breaths of summer. Soon the smells of cut grass, sweat, bug spray, and grilling meats from neighbor's barbecues will be replaced by crisp inhalations of wood smoke and dead leaves.

I turn Merlin up the dirt road that leads us on a four-mile round trip through forest, field, and streamside. Back home there's a loaf of wheat bread on its first rise in a Pyrex bowl and a stainless steel saucepan of raw goat's milk turning into chèvre. I am getting only a half-gallon every 36 hours now, but it is enough to keep me in milk for my cereal and cheese. I think Bonita will be bred this month and so will my new goat Francis, purchased to keep Bonita company and double my milk production!

I mull all of this as the trail gets steeper, leaning
forward as my horse climbs. The road levels out and
I lean back. These motions are now as normal and
thoughtless as putting on pants.

Merlin doesn't want to climb the steepest part of the
dirt road and I know him, and myself, well enough
now to handle his fit. He stops and turns around to
trot home and I tsk-tsk and smile. In a split-second
I loosen my right rein, gently pull my left, and
spin him in three circles until he stops on his own.
I turn him uphill, offer my heel, and loosen the
reins. He doesn't budge, just turns back around to
go to the farm. I spin him again and this time when
we're facing uphill I loosen the reins and kick at the
same moment. When he balks again I use the over/
under rope on his horn — it has a piece of rawhide
on the end. I send a light flick back to his rump and
he canters up the steep dirt slope.

He has learned, finally, that I am more stubborn
than he is.

6 September. *Putting Up*

I spend most of the day in the kitchen. Outside it's overcast, building for storms, but inside the house is thumping with good smells and work. I am canning strawberry jam and dill pickles with Cathy Daughton and her three boys, who have driven up from their stead in White Creek to help out and stay for dinner.

I have made plenty of jam in my day, but this is my first time making dill pickles in the water bath canner. I thank the people at Ball for a premixed dill pickle packet I picked up at the IGA for two dollars. I pour it into a saucepan with 6 cups of water and 2½ cups of white vinegar and bring it to a boil. We pour it over fresh-cut cucumber spears, seal the jars with the metal rings, and let them boil for fifteen minutes.

That's it. Mix, boil, cover, seal, and can. And then of course, on some December night, devour.

9 September. *Local Sheepdog Trials!*

It is chilly this morning, and the green grass on
the hillside is tipped with cold dew. It's not a frost,
not by 20 degrees, but the 50-degree morning is
a wakeup call. It's only September. October will
bring many mornings that start with a warm stove
and nights that end with it, but during the day those
fires will go out. Before you know it, though, stoking
those stoves will become my *real* full-time job. My
life will revolve around words and fire. A primitive
and happy combination. Just as I like it.

Today I'm off to spend the day at the Merck Forest
Sheepdog Trials. I will watch, help keep score if
asked, and catch up with some old friends. It's
something I have missed dearly, the lessons and the
excitement of the trials. The outdoor chores are
done and coffee is on the stove, so all I need to do is
put on a kilt and some rubber boots and hit the
road with Gibson. Time to attend a sheepdog trial!
Away to me!

10 September. On Draft

Today is Merlin's debut, his first time driving in
public. We (Washington County Draft Animal
Association) all meet at the Arlington Grange at
9:30 AM for a pancake breakfast before the ride. For
six dollars we receive a heaping plate of blueberry
pancakes, sausage, potatoes, and biscuits and gravy.
We drink strong coffee and pour Vermont maple
syrup over our flapjacks.

Patty and Steele and Merlin and me. Patty ties a
sunflower and ribbons in Steele's white tail and then
smiles at me. "Now I have something to look at on
the road."

All around us horses are being groomed and fawned
over, harness hames raised over heads and set on
strong backs. I feel so proud to be a member of the
club, so grateful for the blue skies and happy faces.

Steele looks magical and grand, his 1,800 pounds of
muscle and energy, tipped with the sunflower, ready
for an oil painting. Merlin has a single gray goose
feather tied in his black mane. Before I know it, we
are there among the big horses and wagons, awaiting
our turn to join the parade.

"This is it, M," I say, quietly so no one else can hear. "Do your best, be safe, know how much I love you, you big lug." I ask him to walk, and he does as I ask, as I know he will. He is as smooth and calm as can be, just rolling along the river road. I feel like Gandalf in his pony cart. How did I get here? How the hell am I lucky enough to be out with a beautiful Celtic pony on a sunny autumn day in a smart-looking cart? I'm not a heavy load for a Fell pony, but I can't imagine how he can haul that much gratitude for eight miles. It must weigh 20 stone, at least.

We ride along River Road for four miles, just Merlin and I. Then we fall in line a few carts behind Patty and Steele. Merlin's crinkled tail swishes and his ears flick back and forth listening to the bells and trotting hooves all around us. I look across the river to Route 313, the road I took to work every weekday not long ago. It is busy, cars rushing at a clip our horses could never match. To watch that from a pony cart on a dirt road is pretty darn neat, and sobering — like reading an obituary of a life I once led.

When we take a break in a small field near the town of Arlington, I spot familiar faces — Phil Monahan and his daughter Claire. I invite Claire to ride back with Merlin and me and she literally jumps up and down. I have my first passenger, a second-grader.

We join the faster group for the ride back and make the four-mile trip home in about thirty minutes! Merlin is sweating now, but just. He is in the best shape of his life this summer, and it shows. Claire is great company and mighty brave. She helps me with Merlin's tack afterward and gets water for his bucket.

With the teams back, the sun warm, and appetites awake, we head into the Grange to do what we members of the WCDAA do best: eat. We fill plates once again, this time with chowders and buttered bread, mac-n-cheese and meatballs, and all sorts of cakes and desserts. Everyone, passenger to teamster, seems thrilled with the event. Nothing went wrong, the weather is perfect, and the food as plentiful as heaven's own rain.

15 September. *Song Dogs*

No one who raises livestock can avoid predators. My mountain teems with foxes, hawks, opossums, raccoons, weasels, fishers, and coyotes. Of all these, it's the fox I disdain the most. The others have the decency to come in the dark of night or at least act guilty. But the foxes are bold as brush strokes. They walk right up to a young free-range pullet in broad daylight while I'm hoeing in the garden nearby, and they're leaping into the brush with it before I can move a muscle.

Which is why I love coyotes. The song dogs are my favorite of all the local carnivores. I love their yips and howls at midnight. I love their wit and proud stance in a field at sunset. Coyotes are large, thick-boned, and heavily coated around here, beautiful in a way only a predator in its prime can be. They are masters of adaptation and still wary of dogs and humans. No coyote has ever taken so much as a chick off this farm, probably because my animals live close to my house and the perimeter of my land is rank with the smell of domestic dog pee.

But what I love most is that when coyotes take over a territory, they make sure every fox is on the run, by chasing, killing, or threatening it out. So as I fall asleep to the song dogs' chorus, I just smile. The sheep and the birds are safe, because the neighborhood watch is on duty.

17 September. *The Bad Part*

This morning I am feeding the sheep near their
shed and spreading straw inside for clean bedding
(rain all day today so I want them comfortable
inside) when I back into a paper wasps' nest sneakily
built in the interior walls.

The bad part: I am riddled with stings.

The good part: There is no good part.
They are wasps.

HOLY·OCTOBER

On a chilly morning in mid-October, I am awake in a way my mind and body do not understand at any other time of the year. I curl my spine, sinking deeper into my quilts, and cannot hide my smile. Gibson's smiling snout meets my face, and sloppy dog kisses greet my day. I grab him and pull him close and scratch his belly. His tail thumps on the mattress, and he squirms in delight.

I inhale a potpourri of smells: my dog, maple leaves, and wood smoke from a neighbor's stove. I purposely leave my windows open, inviting in the 35-degree chill of the blue dawn. The house isn't cold, but it is brisk. It would be freezing if it weren't for the fire still blazing in the woodstove. The combination of indoor firelight and open windows makes my living room feel like a campsite.

11 October. *Hero*

Tonight my small 1850s farmhouse is lit by candles,
along with some bright lamps for reading. I can
hear the blustery rustle of the dead leaves outside
and smell that swirl of decay as I nestle by the
fire. As a resident of the Hudson Valley, I choose
as my evening companion a hardcover edition
of Washington Irving's *The Legend of Sleepy Hollow*.
One passage in particular strikes a chord:

*I profess not to know how women's hearts are wooed and won.
To me they have always been matters of riddle and admiration.
Some seem to have but one vulnerable point, or door of access;
while others have a thousand avenues, and may be captured
in a thousand different ways. It is a great triumph of skill to gain
the former, but a still greater proof of generalship to maintain
possession of the latter, for a man must battle for his fortress
at every door and window. He who wins a thousand common
hearts is therefore entitled to some renown; but he who keeps
undisputed sway over the heart of a coquette is indeed a hero.*

If I'm anything like Ichabod's coquette,
then October is my hero. I'm a confirmed
bachelorette. I don't date for sport, and
I don't waste any time on men I can't love
the way I love this sweet, holy month.
Few men have made me feel the way October
mornings do, and those men have been
forgotten, their faces just parts of photographs
instead of the topography I once touched.

But October, I welcome him with all I've got.
His wildness thrills me. His color warms me from
the inside out. Those oranges and reds that cover
his dirty brown skin, his voice deep, his time dear.
He's the love of my life.

A girl could do worse.

14 October. *Winter Hive*

The bees are more ready for winter than I am.
Their largest box, at the base of their stack of
supers, is loaded with honey and brood. No matter
what happens they won't be going hungry. Their
kind of foresight would be both rare and enviable
in humans. What would they think of me, waiting
until summer to start my woodpile? Not much,
I bet.

Some folks prepare their bees for winter by
wrapping the hive in light insulation, wool
blankets, or Styrofoam. I do not. My hives are
in a well-protected area with their backs to a
mountain ridge and tree line and their sides
hugged by a honeysuckle. As long as no snow
blocks their ventilation and exits, they should be
just fine. It'll be a long winter for these working
women, but at least they have a safe place to
stay with plenty of eats.

18 October. *Four Strings*

A fiddle is just four strings held by tension over
a wooden box with holes in it. You need only four
basic finger positions to get started, and those
same positions are the same on every string. In
about an hour you can learn the entire map of the
instrument, and in a few months you won't be
able to imagine your life without it.

You'll strap your fiddle over your back with baling
twine to go to campfires and friends' BBQs. And
when you start a raucous round of "Old Joe Clark,"
everyone will be shocked you had it in you, but you
won't be. It's a natural outcome of practice and love.

Musical instruments are like gardens. Anyone who
can follow basic instructions and access sunlight,
soil, and seeds can grow a patch of lettuce greens.
In the same way, anyone with a tuned fiddle, a
chunk of rosin, and some determined effort can
grow a song. It just takes learning new moves,
understanding a new animal. And like gardens you
can be as simple or complicated and ambitious as
you want to be.

22 October. *Planning Parenthood*

Lots of plans ahead for the animals of my homestead —
starting with the Testosterone Reduction Program.
There are just too many roosters here at CAF, so
I'm taking a crate of seven of them to be processed.
Three of those boys are Freedom Rangers that escaped
capture during summer slaughter, but the rest are
just accidental hatchery misfits that happened to be
male in a very female workforce.

All of them will make Freezer Camp by Tuesday
night. That'll reduce the farm to a few choice males
and a happy group of mixed-breed hens.

The goats are bred and back in their pen with Monday,
the ram lamb. Kidding will begin around March 9
and last all that week, and while I'm not nervous about
my first goat, Bonita, first-time mom Francis might
need midwifery help.

Honey bees are important and valuable citizens of
Cold Antler. I'll order a new colony for spring.

The piglets have stopped escaping and remain together in their new deluxe pig pen. In the morning I go into the pen to serve up breakfast and they are always spooning together in clean hay, snoring. I feel as if I've invaded their personal space, but they care little when last night's roasted chicken drippings and some cracked eggs coat their grain. All is forgiven in food.

31 October. *Hallowed Day*

I greet my cold morning with love. I light the stove, feed the animals, and decide to eat out on this bright morning. I cook up a breakfast burrito of my hens' eggs, a home-grown tomato, and smoked cheddar and wrap it in foil. I pour coffee into a Thermos and make sure to stash some of Merlin's favorite cookies.

I ride up the mountain to a view of the valley below, taking in the shock of color and morning light. I stand there savoring the scene and the burrito, and I venerate the great and magical and holy day I pine for all year: Halloween. The anchor of so many agrarians for thousands of years and the oldest holiday that has survived to these modern times.

Halloween! That old girl has been celebrated for something like 6,000 years and is still going strong. It is different now, of course, but I'm doing my part to bring back a different sort of observation, one honoring the land. It's the least I could do for the month I love.

HALLOWS

The celebration that became known
as Halloween is believed to have
originated around 4,000 BCE, making
it the oldest holiday we still have.
Halloween is about the big things:
Life and Death. It's a day to be quiet
and realize that this is all over
too soon and to celebrate what's
left with all you've got. Only when
you accept your own mortality are you
truly free to live this life without
hindrance or fear.

Halloween is an affirmation of the
gift of life.

EPILOGUE

I follow my year as a circle. Halloween, Shearing
Day, Trail Ride Sunsets: these are my new Days
of Grace. They are the tradition I want to hold dear.

These new holidays of my farm life are not
replacing the holidays of my childhood. Christmas
and Thanksgiving still have red circles on the
wall calendar and make me smile. But this new
grace has replaced the emotions I once felt on
those days. You can't wait in line at the mall to sit
on the Great Pumpkin's lap, kids. You need to
walk out past the flocking crows and find him.
The oldest religion in the world is out where
the sun sets and rises and seeds come up to start
the story all over again.

The longer I dedicate myself to this observational,
spiritual farming, the more I realize certain things
to be true: such as, you can't run a small farm
and not be religious. It won't work. I don't mean

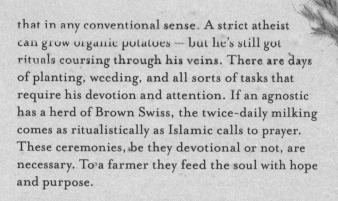

that in any conventional sense. A strict atheist can grow organic potatoes — but he's still got rituals coursing through his veins. There are days of planting, weeding, and all sorts of tasks that require his devotion and attention. If an agnostic has a herd of Brown Swiss, the twice-daily milking comes as ritualistically as Islamic calls to prayer. These ceremonies, be they devotional or not, are necessary. To a farmer they feed the soul with hope and purpose.

So now I live this life, one shaped daily by the tasks and rituals of raising my own vegetables and animals. My morning prayers no longer slide up and down a string of beads but instead take the form of pouring 5-gallon plastic buckets of water into troughs. I start the day with hay bales, grain scoops, and the ardent observation of my furred and feathered congregation. Homilies come in low bleats and soft clucks. Chores become acts of

blessing. My daily work is always heading toward a larger holiday: shearing, lambing, and bringing in this year's ram. And it has all given me the gift of purpose in the most practical sense: Do this work and it will sustain you. Do this work and you are alive.

This newfound religion is a faith of constant motion. It is alive and on the hunt, panting and begging for just a few more kisses, a few more sunsets, a few more winters with a larder full of hope. It spends its Sunday mornings loping past those classical buildings with Tiffany windows, running through the tall grass beyond the graveyards and diving deep into the corn.

You can't obtain salvation for your sins out there, but you can come to a quiet understanding of how little most of them matter. You can't focus very well on original sin when your heart is soaring under a million stars. You will not get the answers to the meaning of life, but you won't go home hungry.

Perhaps that is a good place to start.

Let the wealthy and great

Roll in splendour and state,

I envy them not, I declare it.

I eat my own lamb,

My own chickens and ham,

I shear my own fleece and I wear it.

I have lawns, I have bowers,

I have fruits, I have flowers,

The lark is my morning alarmer.

So jolly boys now,

Here's to God Speed the Plough

Long life and success to the Farmer!

— "GOD SPEED THE PLOUGH"
ENGLISH FOLK SONG